ELEMENTS OF GEOLOGICAL MAPS

FOR

STUDENTS OF GEOLOGY, GEOGRAPHY & CIVIL ENGINEERING

SECOND EDITION

S.K. Chadha
M.Sc., Ph.D.
Professor & Head, Deptt. of Geology,
Govt. G.M. Science College,
Jammu Tawi (J & K State), India

CBS

CBS Publishers & Distributors Pvt. Ltd.

New Delhi • Bengaluru • Chennai • Kochi • Kolkata • Mumbai

Bhubaneswar • Hyderabad • Jharkhand • Nagpur • Patna • Pune • Uttarakhand • Dhaka

ISBN: 81-239-0372-3

First Edition: 1995
Reprint: 2006, 2009, 2010, 2012, 2014, 2015, 2020

Published by **Satish Kumar Jain** and produced by **Varun Jain** for
CBS Publishers & Distributors Pvt. Ltd.,
4819/XI Prahlad Street, 24 Ansari Road, Daryaganj, New Delhi - 110002
delhi@cbspd.com, cbspubs@airtelmail.in • www.cbspd.com
Ph.: 23289259, 23266861, 23266867 • Fax: 011-23243014

Corporate Office: 204 FIE, Industrial Area, Patparganj, Delhi - 110 092
Ph: 49344934 • Fax: 011-49344935
E-mail: publishing@cbspd.com • publicity@cbspd.com

Branches:
- ***Bengaluru:*** 2975, 17th Cross, K.R. Road, Bansankari 2nd Stage, Bengaluru - 70 • Ph: +91-80-26771678/79 • Fax: +91-80-26771680
- ***Chennai:*** No. 7, Subbaraya Street, Shenoy Nagar, Chennai - 600030 Ph: +91-44-26681266, 26680620 • Fax: +91-44-42032115
- ***Kochi:*** Ashana House, 39/1904, A.M. Thomas Road, Valanjambalam, Ernakulum, Kochi • Ph: +91-484-4059061-65 Fax: +91-484-4059065 • E-mail: cochin@cbspd.com
- ***Kolkata:*** 6-B, Ground Floor, Rameshwar Shaw Road, Kolkata - 700014 Ph: +91-33-22891126/7/8 • E-mail: kolkata@cbspd.com
- ***Mumbai:*** 83-C, Dr. E. Moses Road, Worli, Mumbai - 400018 Ph: +91-9833017933, 022-24902340/41 • E-mail: mumbai@cbspd.com

Representatives:
- Hyderabad: 0-9885175004
- Patna: 0-9334159340
- Jharkhand: 0-9811541605
- Nagpur: 0-9021734563
- Pune: 0-9623451994
- Uttarakhand: 0-9716462459

Printed at: J.S. Offset Printers, Delhi (India)

A token of
life—long
Homage
and
Affection
to my
grandmother

SMT. SULAKHANI DEVI

PREFACE TO THE SECOND EDITION

The book has been revised and made uptodate to meet the requirements of students of geo-sciences and applied sciences of different universities of India. Errors, omissions and suggestions, brought to my notice, will be thankfully acknowledged and incorporated in the next edition.

S.K. Chadha

PREFACE TO THE FIRST EDITION

It is crystal clear that the students of Geology, Geography and civil engineering are greatly handicapped due to non-availability of any concise, comperhensive, simple and cheap book on the subject. The present book is therefore, designed to offer all these things. A new approach designed to help the students working with little or no supervision is employed; each topic is simply explained and illustrated by figures, and more than thirty four exercises upon the geological maps have been included, so as to make the students to attempt and understand the subject in a proper and a better way.

In bringing out this book, I do not claim any originality. It is only due to the inspiration, co-operation, and encouragement of my friends and teachers that I have been able to write these pages. It is my proud privilege to extend my hearty thanks to Dr; G. W. Chiplonkar (Poona); Dr; N. Rajagopalan (Annamalainagar); Prof: T. N. Muthuswami (Madurai); Dr; A. Sriramadas; Prof: P. K. Mukherjee (Calcutta); Shri Jagadishwar Johri (Dharmsala); Dr; W D. West (Saugar); Prof: V. Subramanyam; Dr; N. Dasarathi (Jammu) Dr; F. G. H. Blyth (London); Dr; G. M. Bennison (Birmingham—England); and the publishers M/s Penguin Book Ltd; Middlesex (U. K.); M/s Dastane Ramchandra & Co; Poona; M/s. George Allen & Unwin (P) Ltd: London; M/s. Edward Arnold (Publishers) Ltd, London; M/s. The World Press (P) Ltd; Calcutta, Shri K. L. Bhan; Jammu, and M/s V. M Studios; Jammu; for their kindly permitting me to re produce matter from their publications and supplying other material for my book

My sincerest thanks are also due to my respected teacher Dr; S. N Wakhaloo, head of the post graduate department of Geology, J&K University, Jammu, for his kindly going through the manuscript and his valuable suggestions for the improvement of the book.

Suggestions for its further improvement will be thankfully received.

S. K. Chadha

CONTENTS

Chapter		Page
	Preface	
1	Introduction	1
2	Representation of Relief	3
3	Conventional Symbols	5
4	Dip and Strike	7
5	Study of Outcrops	11
6	Conformity and Unconformity	17
7	Faults	21
8	Folding	25
9	Igneous and Metamorphic Rocks ...	29
10	Interpretation of Geological Maps ...	31
11	Construction of Structure Section and Geological History of the Map ...	33
12	Exercises upon Geological Maps ...	35–53
	Appendices ...	55–59

1

Introduction

Fig. I :Representation of globe on a paper.

A map is a representation of a part or whole of the globe on a plane paper. Map is always drawn to a particular scale. In it different details are incorporated, so as to make it useful for specific purposes.

Geological Map

Maps are of different kinds, but a geological map is essentially a plan exhibiting the outcrops and structural characters of the various geological formations in the area represented. In its construction the help of the topographic map is also required.

Types of Geological Maps

The maps studied in geology are of four types, which are explained as under :—

(a) Structural Maps (b) Outcrop Maps (c) Areal Maps and (d) Surficial Maps.

(a) Structural maps are easy to prepare, especially when the rock units are shown by symbols in black and white. In such maps, different methods are employed to represent the geological structures. The two most commonly used methods are "structural symbols" and "structural contours".

(b) Outcrop maps represent the areas where the bed rocks are better exposed. The different beds are recognised by means of symbols

(c) Areal maps are plan views of the geology. They represent both the observed distribution of the geological units as well as the inferred distribution of the "Solid Geology" and Vegetation, as if the latter were swept entirely away.

(d) Surficial maps exclusively show the distribution and character of various kinds of surficial materials.

Scale of a Map

The scale of a map is defined as the proportion between the distance on the ground and the corresponding distance on the map. If we say that the scale of a map is 1″ : 1000′, it means that the two points which are 1 inch apart on the map, are actually 1000 feet apart in the field The most commonly used scales are :—

1" : 1000'; 1" : 800'; 1" : 500'; 1" : 1 Mile; 1" : 2 Miles; 1" : 4 Miles; 6" : 1 Mile 1 cm to 200 m; 1 cm=100 m; 1 : 50 000; 1 : 20,000; 1 : 10,000 etc.

Significance of Geological Maps

The geological maps are of unique importance to study the topograhical features of the earth, the distribution of rocks in general, the structural behaviour of mineral deposits, the relationship of the rocks, their order of superposition, age and their mode of formation. They also portray the extent and character of the strata. The succession and structural peculiarities often help to locate the concealed deposits of economic value. The study of such maps also provide clues to the possible sites for tunnels, dams and other costly engineering projects.

Thus geological maps are very useful to mining engineers, civil engineers, economists and to geologists.

1(a) The scales expressed in Representative Fractions on the geological maps, can conveniently be expressed in other equivalent scales in centimetres to metres and in inches to feet (or miles), for instance :—

1 : 50,000=1	cm to	500	metres	=1	inch to	0·788 miles	approximately.
1 : 40,000=1	,,	400	,,	=1	,,	0·63 ,,	,,
1 : 30,000=1	,,	300	,,	=1	,,	½ ,,	,,
1 : 20,000=1	,,	200	,,	=1	,,	1680 feet	,,
1 : 10,000=1	,,	100	,,	=1	,,	840 ,,	,,
1 : 5,000=1	,,	50	,,	=1	,,	420 ,,	,,

(b) Conversions should be made as under :—

1 Mile=1·61 Kilometres	OR	1 Kilometre=0·62 Miles
1 Feet=0·30 Metres	OR	1 Metre =3·28 Feet.

2

Representation of Relief

The term "relief" may be defined as the configuration or uneven nature of the earth's crust. It reveals the nature of the land surface, whether plain or hilly. The maps of the plains are drawn without any difficulty. Different methods are being employed for representing the relief but none is completely satisfactory. In certain maps, two or more methods are combined so as to offer a better representation of relief. Howsoever the most modern methods for exhibiting the relief and elevations are as under :—

(i) Spot Heights (ii) Hill Shading (iii) Hachures and (iv) Contours.

Spot Heights :—These are meant to depict the exact elevations of points above the sea level. Alone, they do not give fair idea about relief but are very useful in indicating the heights of hill tops. These are represented by dots, followed by the height above the sea level.

Hill Shading : This method gives a graphic but more generalised idea of the relief. It is meant for maps of vast areas on a smaller scale. Here the steep slopes are represented by dark shades, while the gentle slopes by light shades. The hill shading neither shows exact heights nor adequately the relative altitudes and gradients.

Hachures :—These are short lines of shading drawn on a map to show differences in the slope of the ground. If the slope is steep the hachures are close together and thick, if the slope is gentle they are farther apart and thinner. In the hachured maps, the darkest portions represent very steep slopes while the lighter shades represent very gentle slopes. The white parts of the hachured map represent level mountain tops or level valley bottoms. Such a map, therefore, gives a fairly good impression of the configuration of the land. The hachures are drawn from the highest point of a mountain to the foot, and these must be perpendicular to the contour lines. The hachures do not give any clue of the actual height of the land above the sea level. In special cases, the hachures are drawn on the contoured maps to represent the minor features of the relief, often omitted by contours. This is done only when the vertical interval is great. This method was invented by Lehmann by the close of the 18th century.

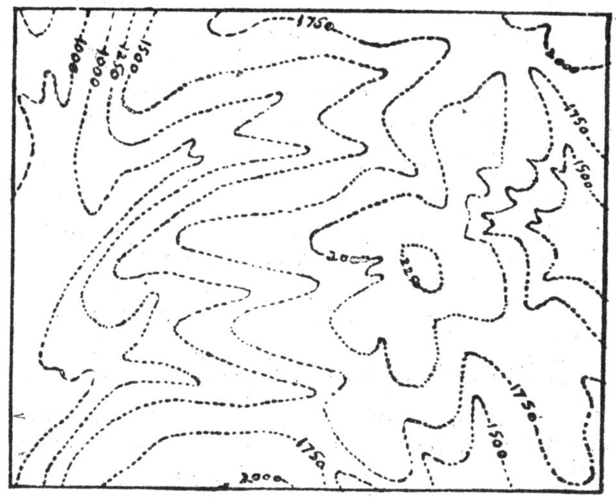

Fig. 2 : Showing Contours

Contours :—This is the ultra modern method of representing the relief. A contour may be defined as the representation, on a map, of an imaginary line joining all points of the same height above the sea level ; and at a given vertical interval. The height is represented by a figure, marked on the contour lines. A contour line on a map marked

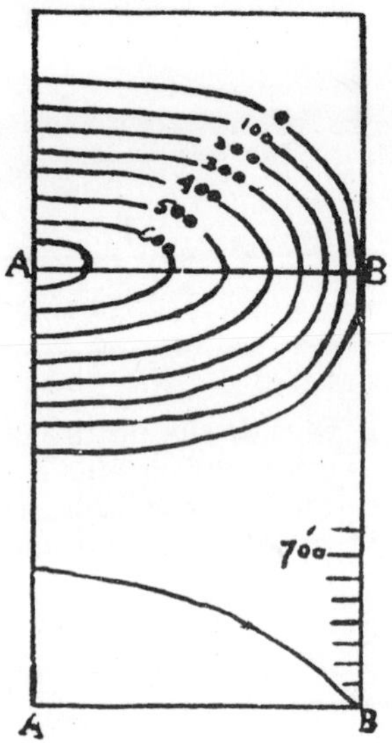

Fig. 3 : Convex Slope

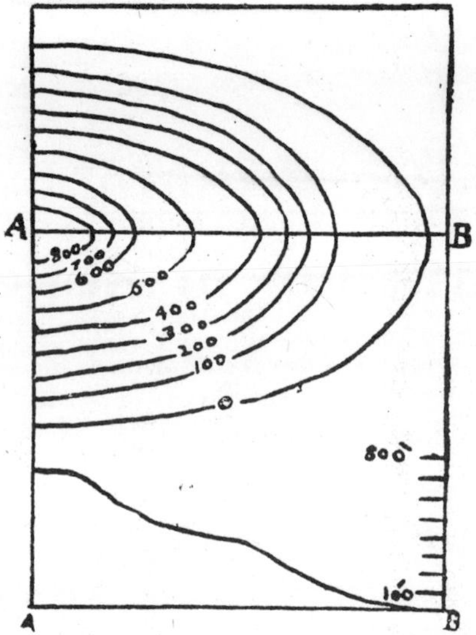

Fig. 4 : Gentle & Steep Slope (irregular slope)

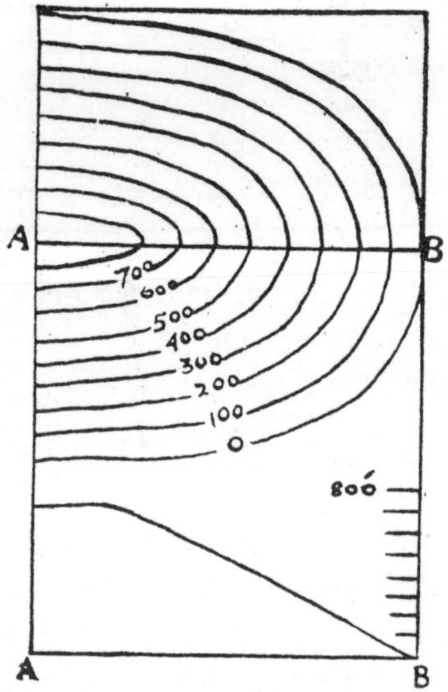

Fig. 5 : Uniform Slope

with a figure 250 feet means that all places that are connected by it are situated at the height of 250 feet above the sea level.

Characteristics of contour lines :—(i) Hill contours are often "U" shaped. The lower contours are enclosed by higher ones and higher by lower ones on the hills.

(ii) Hills and Valleys have contours which go in pairs, the same contour is met twice.

(iii) For gentle slopes they are wider apart and for steep slopes they are close together. The uniform slopes have uniformly spaced contours. The convex slopes have contours which are close to each other, at the bottom of the slopes and wider apart from each other at the top of the slopes. and vice versa, in concave slopes.

Two terms frequently used in connection with the contours are "Vertical Interval" and 'horizontal equivalent' The vertical interval is always constant and usually represented in feet The vertical interval is less for steep slopes and greater for gentle slopes. If the vertical interval on a geological map is 100 feet, the contours will be drawn at 100 feet, 200 feet. 300 feet........300 feet etc. The horizontal equivalent is the horizontal distance between the two successive contour lines. It varies all over the map. The gradient is known by dividing the Vertical Interval by the horizontal equivalent.

In addition to the above mentioned methods, the relief can also be represented by **bench marks.** layer tints etc, but of all these the contours are of unique value.

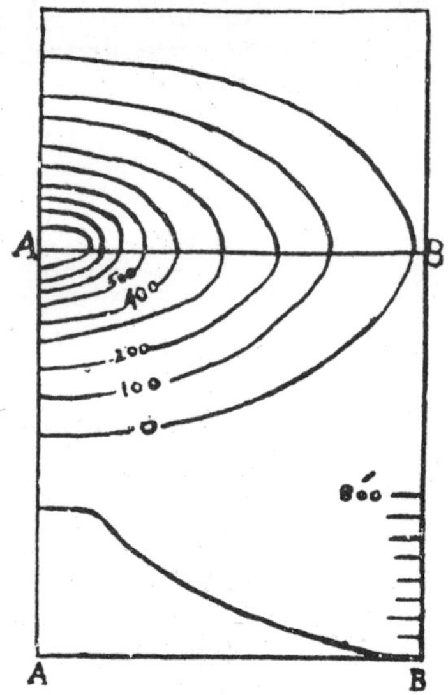

Fig. 6 : Concave Slope

Conventional Symbols

Conventional signs are some specific or definite symbols used for indicating the topographical features on a geological map. The main object of these signs, is to convey into the smallest possible space, the maximum information. These should be simple and self explanatory so as to be easily recognised without a legend. The conventional signs vary according to the scale of the map and the country. The principal conventional signs are generally given at the margins of the maps.

Some of the most important conventional signs and the methods of shading, used in the geological maps are given in the tables No. I to II.

Table I CONVENTIONAL SIGNS

Antiform  The limbs are dipping N-S and the axis E to W.

Bed Vertical 

Bed Horizontal 

Contorted Bed 

Cleavage  The cleavage plane is NE-SW and dipping SE.

Dip Gentle  The arrow indicating the direction of the Dip.

Dip Steep 

Dip at Surface  The Dip is 10° SW.

Dip of cleavage 

Dip of Strata  (Undulating strike)

Dip Undulating 

Fault  Trend of the fault is NE-SW

Fault inferred 

Fault with downthrow direction  The down throw side is towards south.

Geological Junction 

Geological Striae  The arrow pointing the direction of the ice flow

 Igneous intrusion.

Synform  The limbs are dipping E-W and the axis is trending N-S.

Unconformity  Undulating Strata 

Table II Methods of Shading of Rock Types

(i) Igneous Rocks :—

 ⟷ Massive Crystalline Rocks. 

 Granitoid Rocks.

(ii) Sedimentary Rocks :—

 Shale

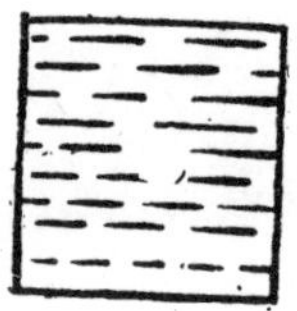 Clay

 Alluvium

 Boulder Clay

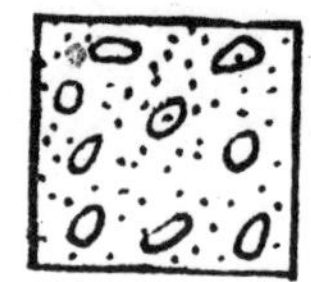 Conglomerate

 Gritstone

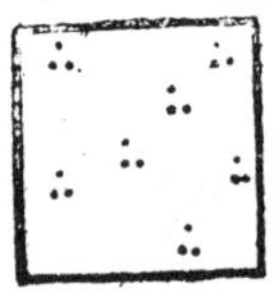 Arkose

 Breccia

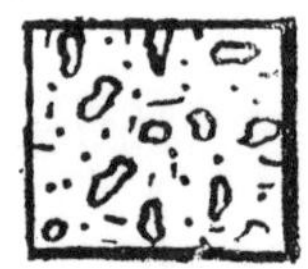 Gravel

 Coal Bed

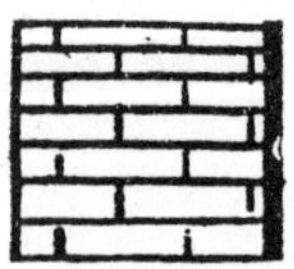 Limestone

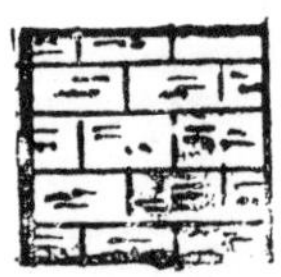 Bedded Limestone

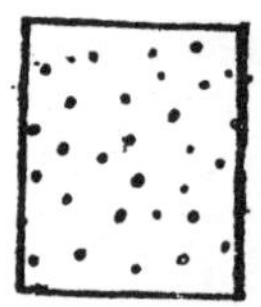 Sandstone

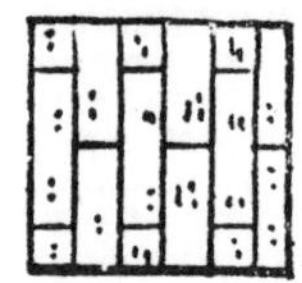 Bedded Sandstone

(iii) Metamorphic Rocks:—

 Schist

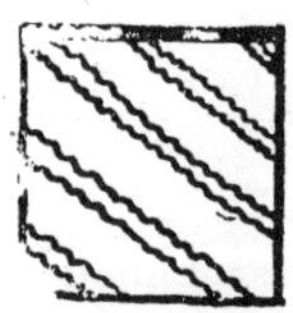 Gneiss

 Slate

Dip and Strike

Dip :— It is of fundamental **importance** in geotectonics and represents geologists' method of deciphering the angle of inclination which the bedding plane makes with a real or imaginary horizontal plane. The *angle of dip* must always be measured at right angle to the strike and is expressed *in degrees*. Dip in any other direction then the strike, is far less and is defined as the *apparent dip*. The apparent dip is usually less than the true dip and is zero along the strike of the inclined plane. The geographic direction in which a bed has a maximum inclination is known as the *direction of dip* The direction of dip at right angles with the strike direction of a surface is always the direction of *true dip*. The *amount of dip* is measured with an instrument known as clinometer, between the inclined surface and the horizontal plane in the direction of dip.

Fig. 47 : Showing the relationship of Dip and Strike.

Fig 48 (a)

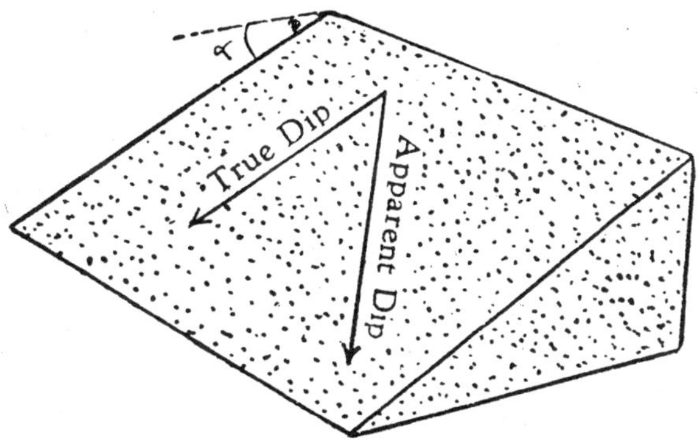

Strike 400′

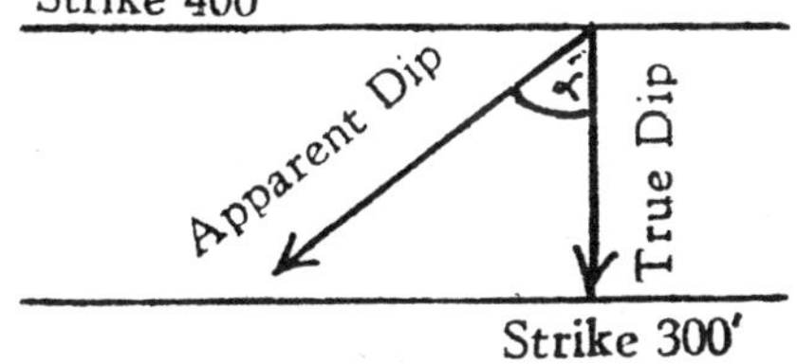

Strike 300′

Fig. 48 (b)

Fig. 48(a) Showing the relationship of true and apparent dip and Fig. 48(b) showing a plan of strike lines to illustrate the relationship of true and apparent dip.

There are three cases of angle of dip.

First-for a horizontal structure, the angle of dip is zero degree.

Second-if the structure is vertical, the angle of dip is 90°. and

Third-if the structure is tilted, the angle of dip is inclined i. e., between 0° to 90°.

Procedure to Determine Amount and Direction of Dip.

The simplest method to determine the dip on a geological map, is to construct a strike line for one surface, then another strike line for the same surface and find out the distance between the two strike lines, by drawing a perpendicular between them as shown in Fig. 49. The direction of the perpendicular from the strike line of higher altitude to the strike line of lower altitude is the direction of dip.

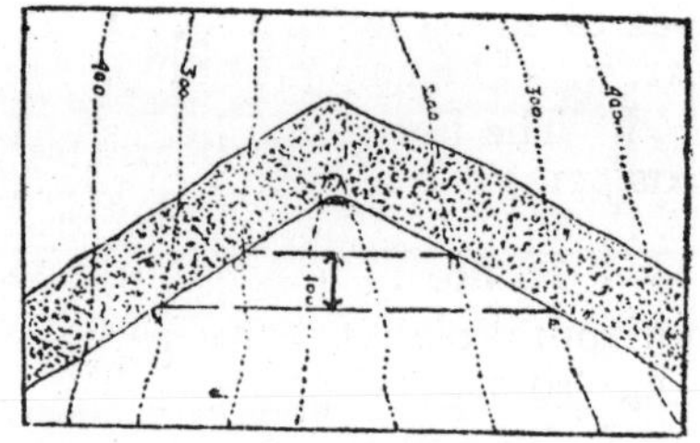

Fig. 49 : Showing the procedure to determine the dip direction.

Also in the Fig. 50 : S_3 is the 300 feet strike line for the boundary of the upper surface of the bed, S_4 is the 200 feet strike line for the same boundary, D is perpendicular distance between the two strike lines. The direction of dip is towards east. In the same way the direction of dip for the other beds can be determined. Now in order to determine the amount of dip from the same map, draw a straight line and measure off a length equal to the perpendicular distance D between the two adjacent strike lines i. e , S_3 and S_4 (or S_1 and S_2). Since the strike interval is 100 feet here, the vertical fall of the bed is 500 feet between S_3 and S_4 (or S_1 and S_2). Take a line equal to 0·5″ (as the scale of the map is 1″ : 1000′) and draw a normal at the right hand side of this line equal to 0·1″ (because 0·1″ represents 100′ w. r. t. the scale 1″ : 100′) Join this end of the normal to the left hand side of the given line. The angle Q thus subtended, gives the amount of dip.

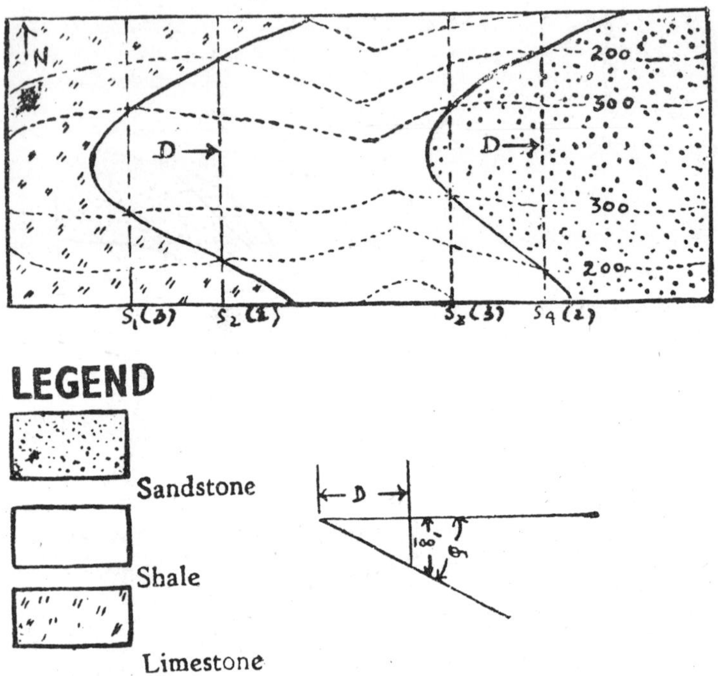

(By courtesy World Press (P.) Ltd.)

Fig. 50 : Showing the method to determine amount and the direction of dip.

There is another "*hit and trial*" method to determine the amount of dip, by applying the formula i. e.

Contour interval / Horizontal equivalent $\times 60^\circ$. In the figure 50, the contour interval is 100 feet and the horizontal equivalent is 500 feet, so the amount of dip is $\frac{100'}{500'} \times 60^\circ = 12^\circ$ approximately. As a rule, it should be remembered that the greater the horizontal equivalent, the lesser will be the dip and the smaller the horizontal equivalent, the greater will be the amount of dip.

The accurate study of the amount and direction of dip leads to the discovery of the actual position of the relief and the detailed history of the circumstances under which those topographical features were evolved in the area

The dip, on the geological map, is indicated by a short arrow. The amount of dip is always indicated in figures by the side of an arrow. In map No. :15, the beds C, D, E, F and G are dipping S-E at an angle of about 22° and it is shown by the dip arrow. The amount of dip refers to the bedding surface at the point of an arrow.

The horizontal bed is indicated on the map by the + sign, while the vertical bed is indicated by —|— sign. The longer axis being the indicator of the strike direction.

Strike :— It is the direction of an imaginary or real line formed by the intersection of a bed with an imaginary or real horizontal planes. Horizontal beds do not have the strike. The strike of a bed is measured in terms of magnetic north i. e.. so many degrees west or east of north.

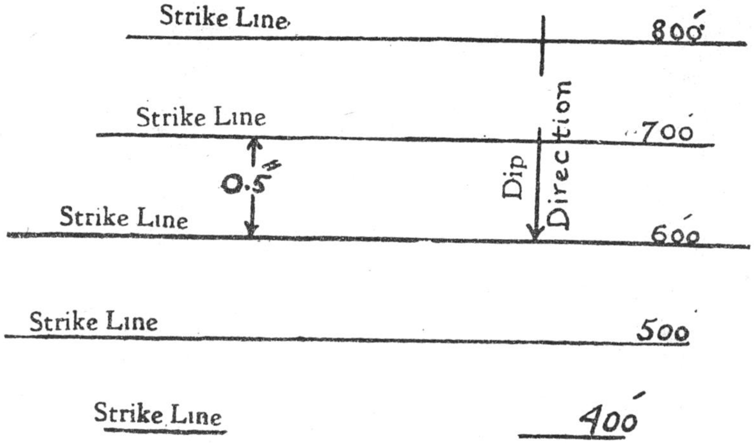

Fig. 51 : Showing a plan of strike lines.

Strike Lines :— Also known as ***Stratum Contours*** or ***Structure Contours***. The relief, as we know, is represented by contours, in a similar way, the contours of a bedding plane are also drawn, which are designated as the structure contours and are parallel to the direction of the strike. In maps, having simply inclined plane surfaces, strike lines will be straight, parallel and if dips are constant, they are equally spaced, as shown in the figure 51. Each contour has its own value.

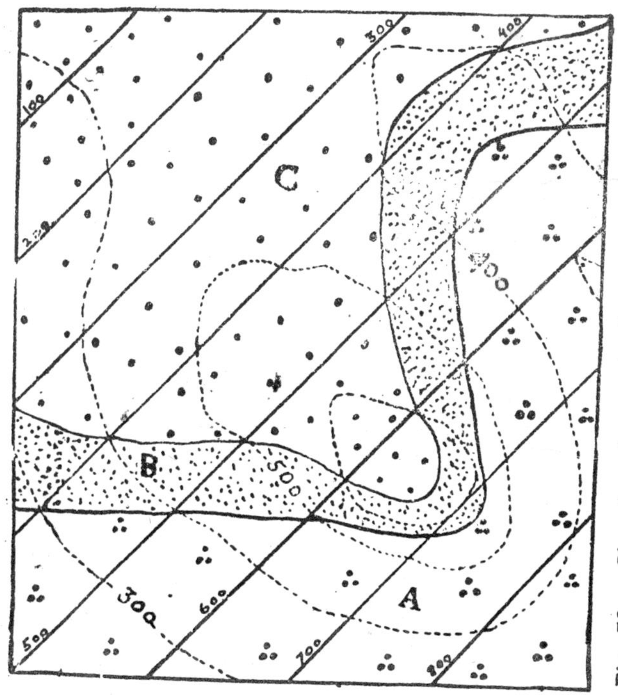

Fig. 52 : Showing the method for drawing strike lines.

Construction of strike lines: —The height of a geological boundary is found where it cuts a contour line; e.g. in Fig 52 The boundary between beds B and A, cuts the 600′ contour at several points. These points lie on the 600′ strike line, which can conveniently be drawn through them. As the beds are simply inclined, the strike lines will, therefore, be straight, parallel and equispaced, as the dips are constant

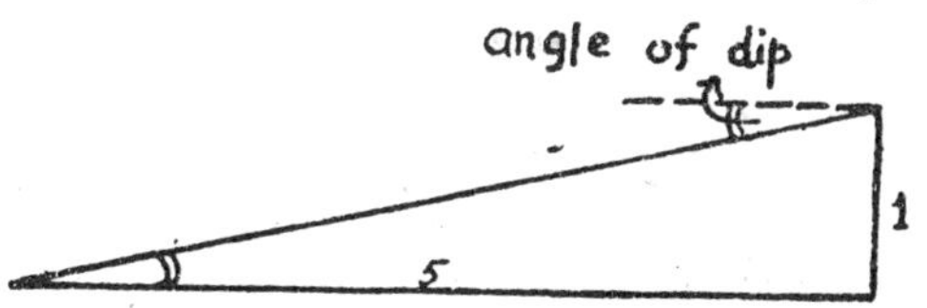

Fig. 53 : Section showing relationship between dip and gradient.

A second strike line can be drawn on the same geological boundary B & A through the points where it cuts the 500 feet contour. From the spacing of the strike lines, we can very easily calculate the gradient. For instance the space between the two strike lines is 0·5″ or 100′ in

500′ (as the scale is 1″ : 1000′), hence the gradient is 1 in 5 as in Fig. 53. It has been noticed that gradients are conveniently used, though the dip is always given as an angle on the map. By the application of formulae of trigonometry, the angle of dip in this case, is that angle which has a tangent of $\frac{1}{5}^\circ$ or 0·2 i. e. $=11\frac{1}{2}^\circ$.

Sometimes, it is difficult to get information about the amount and direction of dip, then we have to depend upon the measurements of the angles of apparent dips in different directions, so as to calculate and compute the direction of true dip of the strata.

---

1. $\pi = 180^\circ \quad 1^\circ = \frac{180^\circ}{\pi} \quad \therefore \frac{1}{5}^\circ = \frac{180^\circ}{\pi} \times \frac{1}{5}$ i. e. $\frac{36^\circ}{\pi}$ then see log tables.

5

Study of Outcrops

An outcrop is the actual area occupied by a certain bed, or a series of beds, or a whole formation on the earth. On a geological map, it is always the projection of this line on to the horizontal plane of the map. The outcrop pattern of the geological

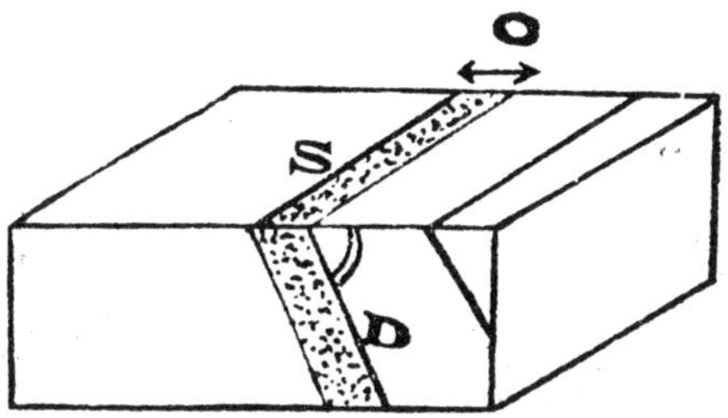

Fig : 54 Showing D=Dip
S=Strike
O=Outcrop.

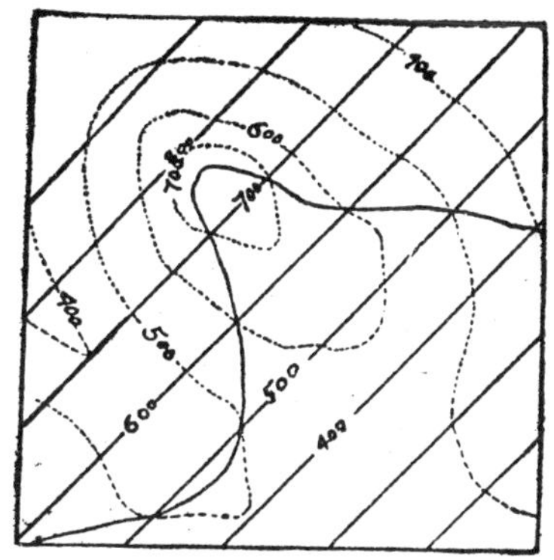

Fig. : 55 Showing an outcrop.

formations of an area is chiefly controlled by the geological structure of the formations and topography. The study of a map commences right from the examination of the outcrops and their attitude. It is by the fabrication of the points of information that the generalised idea regarding the geological structure is known and the geological history of the area is established.

Width of an outcrop :—It is the width of the surface occupied by the bed measured in the direction of dip. The total width of outcrop of dipping beds depend upon three factors. Firstly, the angle of dip ; secondly ; the slope of the topographic surface and thirdly, the thickness of beds. The width of an outcrop, therefore, decreases as the dip increases. The vertical beds have outcrops with a maximum width, while the inclined beds have outcrops with a maximum width. So, in vertical beds, the width of the outcrop is equal to the thickness of the bed, whereas the thickness of an inclined bed is the perpendicular distance between the upper

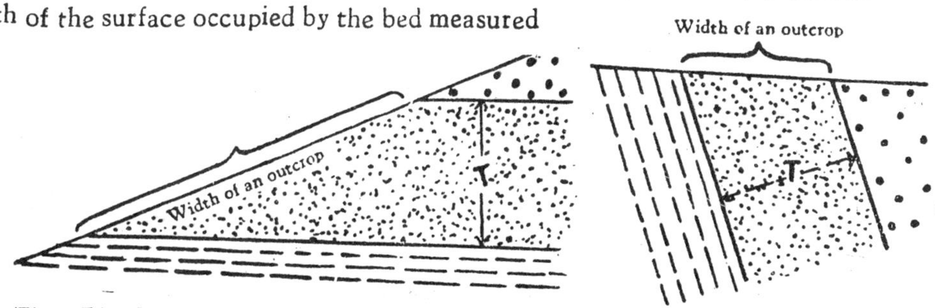

Fig. : 56 Sections illustrating the different widths of outcrop produced by a bed of the same thickness with low and high dips.

and the lower planes of that bed. The vertical distance between these two planes constitutes what is described as VERTICAL THICKNESS. The thickness of a bed outcropping on a ground surface is also equal to the width of the outcrop multiplied by the sine of angle of dip. The vertical thickness, on the other hand, is equal to the width of the outcrop multiplied by tangent of the angle of dip. The other methods to calculate the thickness of a bed are as under:

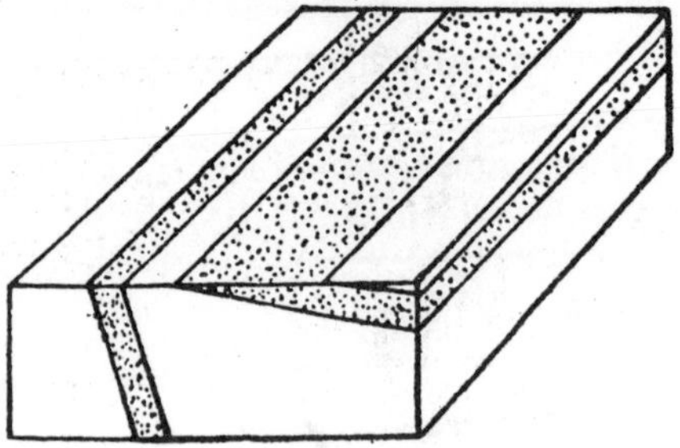

Fig. : 57 Showing the relationship between width of outcrop and angle of dip.

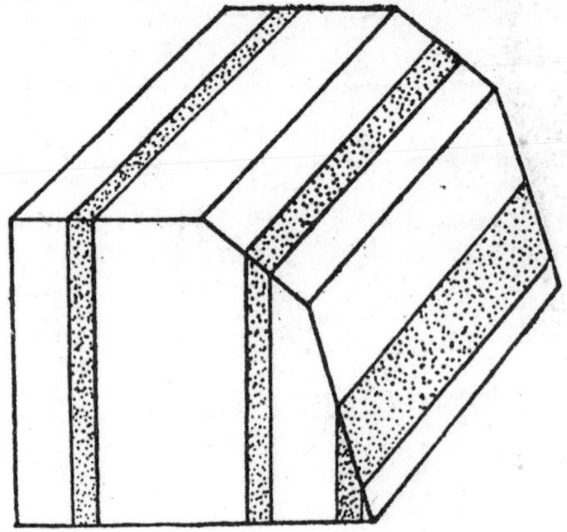

Fig. : 58 Showing the relationship between width of outcrop and topography.

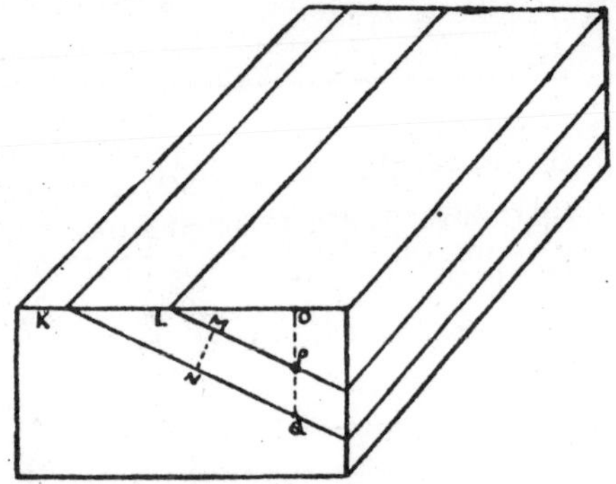

Fig. : 59 Showing the relationship between width of outcrop (K, L), thickness of bed (M, N) and apparent thickness (P, Q).

Dip Method—here section along the maximum dip is taken into consideration and measure the thickness of the bed. Thus the actual thickness of the bed is calculated from the section drawn according to the scale of the map.

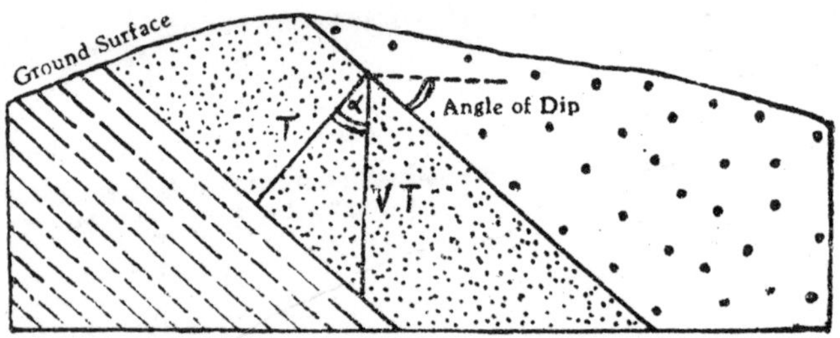

Fig. 60 : Showing the relationship between True and Vertical thickness of a dipping bed.

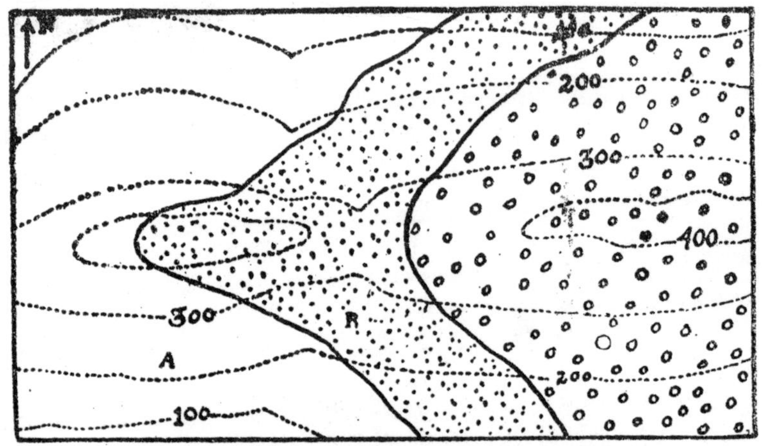

(Courtesy World Press, (P.) Ltd, Calcutta).

Fig. 61 : Showing a map with three beds A, B and C and studying thickness in it by strike method.

Strike Method—here section along the strike direction is considered. The difference in altitudes of the places, where the two surfaces are known to outcrop, would give the vertical thickness of the bed. It can be best studied in the figure 61, if the 300 feet strike line on the boundary surface between B and C be drawn and if the same strike line be further extended to touch the boundary surface between B and A, it is then evident that the thickness of the bed B is equal to 200 feet (300′—100′). As a matter of fact, it is the vertical thickness of the bed B.

Equal Altitude Method—here section is choosen, where the outcrops of the two surfaces of the bed occur at the same altitude, along that line. The character of the intervening surface being of no consideration. It is feasible only when the section along any contour intersecting two surfaces of the bed, is taken.

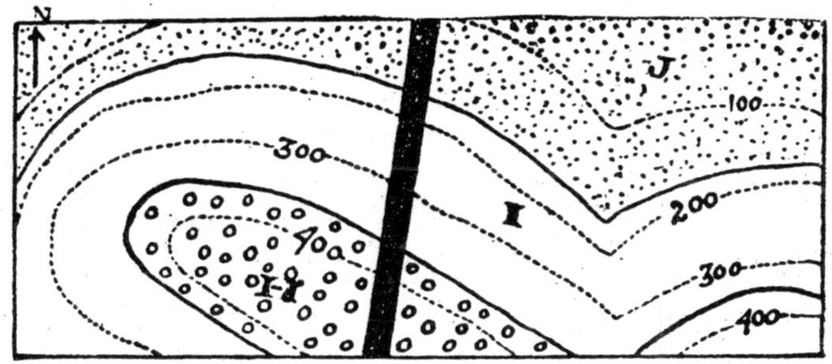

(By Courtesy World Press (P.) Ltd. Calcutta).

Fig 62. Depicting the attitude of the outcrops & determining the thickness of the beds from them.

On scrutinising the maps, it would be observed that if the beds belong to the same formation, the thickness of the uppermost and the lowermost beds cannot easily be ascertained. If the beds are horizontal, the examination of the contours would throw sufficient light on the thickness of the beds. In the figure 62, the bed I is 200 feet thick, while the thickness of the beds H and J can not easily be ascertained, since they are the lowermost and the uppermost beds of the same formation.

Attitude of Outcrops—Outcrops are often plotted on the topographical and geological maps. The different beds are indicated by either colours or by standard symbols. The attitude of the outcrops is of great academic interest. The outcrops of horizontal beds have attitude similar to the course of the surface contours as in Fig. 63. The outcrop while passing from a higher contour to lower contour, often show some amount of apparent dip in that direction. If the bed is surrounded by plane surfaces, the outcrop on a horizontal surface will be enclosed by parallel straight lines, which shall become crenulated as the outcrop is traced on an undulating surface.

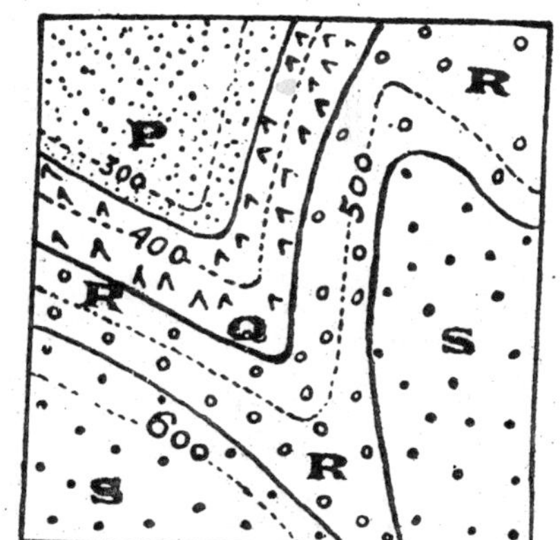

Fig. 63: Showing the attitude of outcrops.

The Rules of V Shaped Outcrops—In case of a valley, where the outcrop is V or U shaped, three cases are possible. First—If the beds dip upstream, the apex of the V will point upstream i. e. in the direction of the dip (fig. 64).

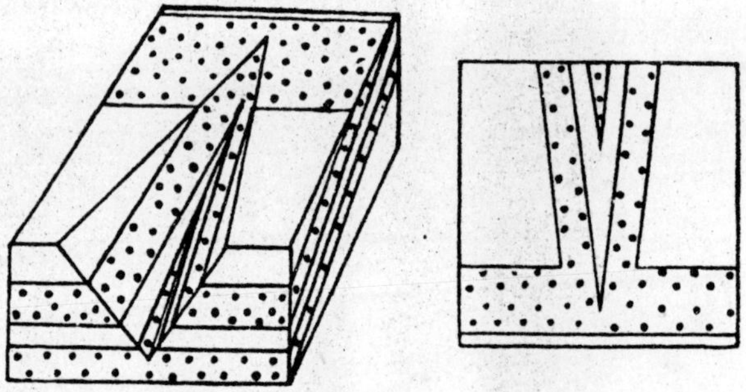

Fig. 64 . Showing shifting of outcrop where beds dip in same direction but more gently than the gradient of the stream.

Fig. 65 . Showing shifting of outcrop where dip of bed greatly exceeds the gradient of the stream.

Second—If the beds dip downstream more steeply than the stream gradient, the V will point downstream, i. e. in the direction of the dip (Fig. 65).

Third — If the beds dip downstream less steeply than the stream gradient, the V of the outcrops is turned upstream i. e. opposite to the direction of dip (Fig. 66).

Fig. 66 . Showing shifting of outcrop, where dip of bed barely exceeds the gradient of the stream.

Repetition of Outcrops—In certain maps, we see the repetition of outcrops, which may be due to number of reasons, such as folding, faulting, prolonged erosion or by dissection of the landmass by streams etc.

Incomplete Outcrops—In order to complete an outcrop on a geological map, find out any two points on the outcrop, where it cuts the same contour line, draw a smooth line from these two points. This line shall be called the "strike line" and has the same value as that of the contour line. Having done this search now for the third point where the outcrop cuts another contour line and from this draw another strike line, which shall certainly be parallel to the first drawn strike line. The second strike line shall have the same value as that of the second contour line. Once the strike lines are drawn, it becomes easier to draw any number of such strike lines, as the vertical interval is known to be uniform. Finally, with free hand, join all such points where each strike line cuts its corresponding contour line. It is thus a simple and appropriate method to complete the incomplete outcrops on the maps.

Determination of altitude with the help of Data—In certain maps, where a contour line does not intersect an outcrop at two places, here the strike line is found from any three points of known elevation on a contact. The three points are joined with each other so as to form a triangle: If the dip is uniform, join the lowest and the highest of the three points. Divide this line into "n" equal parts, where "n" is the difference between the highest and the lowest points, divided by the stratum contour interval. Somewhere along it is a point at the same elevation as the intermediate corner of a triangle. This is the point sought and it can be found either by proportion or by graphic method. From this point draw parallels as necessary, with the same spacing and the required outcrop is drawn by joining the points of intersection of equal topographic and stratum contours.

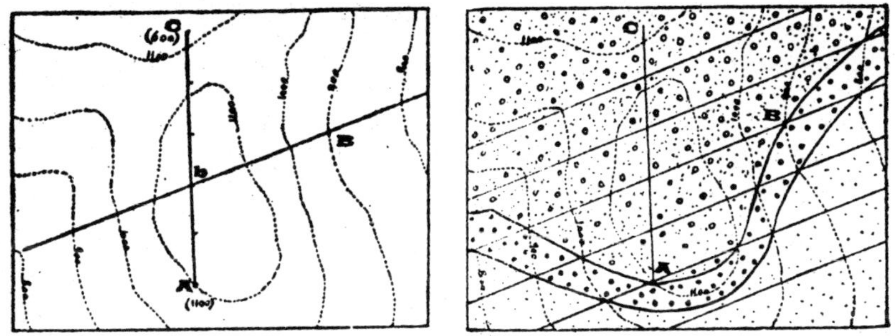

(By Courtesy Edward Arnold (P) Ltd)

Fig 67 . Showing the method to complete the outcrop from by given three points.

EXAMPLE :-

In the Fig. 67, A, B and C are the three points given on the map. Join A and C, being the highest and the lowest of the three points. Divide the line AC into 'n' equal parts, where 'n' is the difference in level between A and C, divided by the stratum contour interval.

So $n = \frac{1100-600}{100}$ $\therefore n = \frac{500}{100} = 5$

One of the points b thus obtained has the same elevation as B and the line B b is, therefore, a **strike line**. Other strike lines can be drawn parallel to B b through the points of division of the AC and be indexed. If necessary parallels can be drawn with the same spacing for values outside this range from C to A. The required outcrop is obtained by joining with free hand, the points of intersection of equal topographic and structural contours as shown in fig. 67. It is how the outcrops are completed and sections are drawn as desired.

Exercise for the Students

Complete the outcrops of the geological Map No: 15 and describe the geology of the area represented on the map.

Conformity and Unconformity On a Map

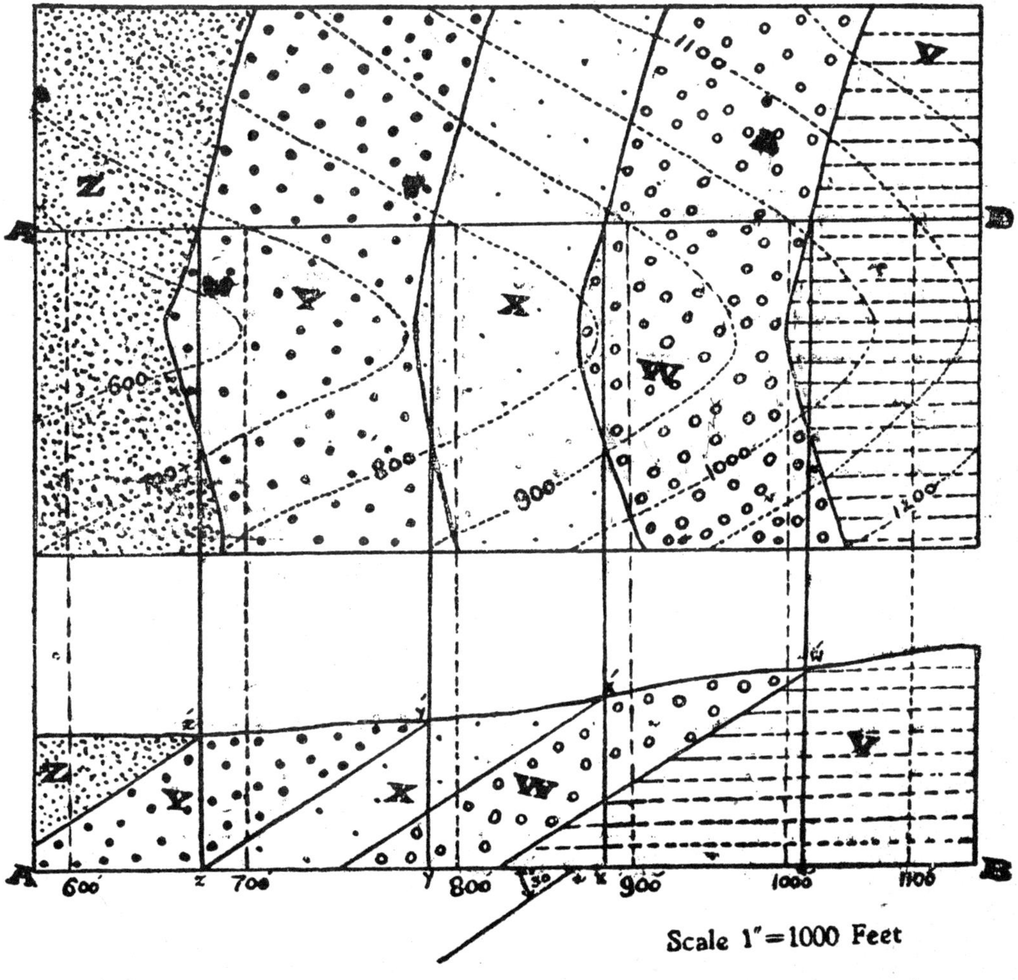

Fig : 68 Showing a map with conformable strata and its section.

The beds on a map may be Conformable or Unconformable. In case the stratified rocks have been deposited layer upon layer and the succession is continuous, the sequence is said to be **conformable**. It is thus obvious that in a conformable strata not isoclinally folded, the lower-most beds are the oldest and those at the top are the youngest in age.

In Fig : 68 the beds VWXYZ are conformable and dip 30° SW.

Procedure to Draw A Section of a Conformable Strata—Make a profile along the line AB by placing a strip of paper on it and plot the points, where the section line cuts the contours. The plotted points are then transferred to the paper on the given base line. With respect to the height of contours, draw perpendiculars from the points. Join with free hand, the end points of all these perpendiculars. Thus profile is ready. Now place again a strip of paper on the line AB and mark the points of intersection of the line AB and the outcrops. Transfer all these points on the base line and erect perpendiculars from ZYXW etc., so as to touch the profile. The points where the outcrops are marked on the profile, thick lines are drawn in accordance with the dip of the beds 30° SW. In the section, the starta is conformable having Uniform dip and strike.

In a conformable sequence, a younger bed comes after an older bed, tcwards the direction of dip. For keeping the beds in order of their antiquity, three points should be remembered.

First—The direction of dip is towards the outcrops of the successively younger beds, if the beds dip opposite to the slope of the ground.

Second—The direction of dip is towards the outcrops of successively younger beds, if the beds dip along the, slope of the ground and the slope of the ground is $<$ than the amount of dip of the beds; and

Third—The direction of dip is towards the outcrops of successively older beds, if the beds dip along the slope of the ground and the slope of the ground is $>$ than the amount of dip of the beds.

Unconformity—In terms of geological history an Unconformity represents a surface of non deposition occuring within a sequence of rocks. During this period, beds already formed may be uplifted or tilted by the crustal movements, which also terminated sedimentation. The tilted strata are thus subject to erosion and are worn down to a great extent before subsidence causes the renewal of sedimentation and the formation of new beds. As a result, one set of strata rest on the eroded surface of an older set of the strata. The break between the two sets of beds represents an Unconformity. The uneven surface of contact between them is known as the **plane of Unconformity**. An Unconformity has three kinds.

1. Disconformity. 2. Non-Conformity and 3. Angular Unconformity.

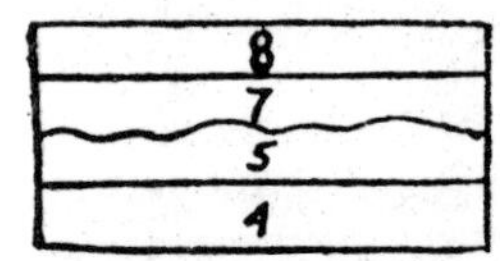

Fig : 69 Showing disconformity

Disconformity—It is an erosional surface among bedded rocks which are parallal to each other.

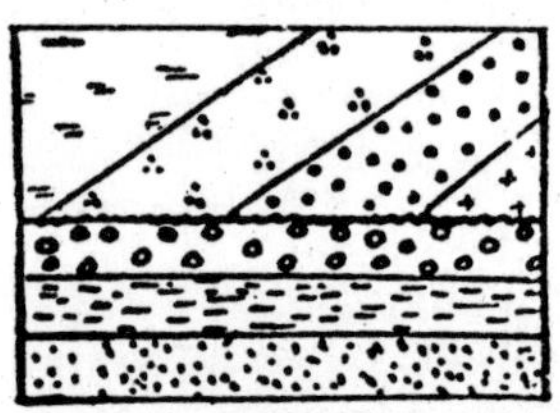

Fig 71 : Showing Angular Unconformity

Non-Conformity—It is an erosional surface between an older massive igneous rocks and the overlying series of younger sediments.

Angular Unconformity—It is an erosional surface which separates two sets of layered rocks in such a way that an angular contact exists between them.

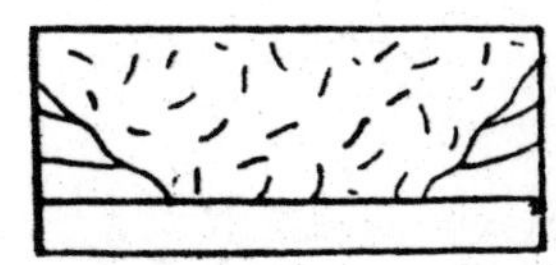

Fig : 70 Showing Non-Conformity.

The phenomenon of Unconformity may be local or regional. It can help in understanding the history of deposition of rocks during the past periods. An Unconformity, in a geological column or succession of beds, is indicated by a zig zag line. In an unconformable strata, it is rarely that the younger strata have the same dip and strike as the older one.

Fig : 72 Showing the formation of an Unconformity.

Criteria to Recognise Unconformity

(i) An unconformity is generally marked by the presence of basal conglomerate

(ii) The lowest beds of the upper set of strata on one side of the Unconformity come in contact with different beds of the lower set of strata on the other side. The older series of beds continue, however, underneath the newer series on the other side of the line of unconformity.

(iii) The beds appear to be trapped up.

(iv) The outcrops often have sharp and angular contact.

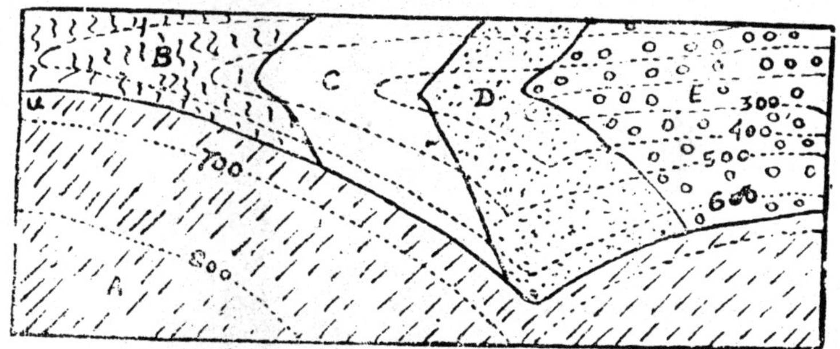

(By courtesy World Press (P) Ltd; Calcutta)

Fig : 73 Illustrating unconformity. The bed A is horizontal, while B,C,D & E dip towards West. The bed A is younger and appear to trap the older beds.

(v) The older beds being truncated by **the younger beds.**

(vi) The outcrops of dykes or faults abruptly terminate against the outcrops of the younger beds

(vii) There is sudden change in the amount and direction of dip of beds; and

(viii) Overstepping of outcrops of the lowest bed of the youngest series of strata over the outcrops of beds of different ages.

Other Structural Features Observed in Maps

Overlap—When the upper beds of a conformable strata occur over a vast area than the lower beds of the same series, the phenomenon is known as an overlap. It depicts, the process of subsidence followed by sedimentation over a gradually extending area and is, therefore, often shown in cases of marked unconformity. On a map, an overlap is often, characterised by a gradual thinning out and finally disappearance of the outcrop of a bed in a Conformable series.

Inlier & Outlier—An outlier is an outcrop which is surrounded by the outcrops of older rocks. On the contrary, an outcrop which is surrounded by the outcrops of younger beds is called an inlier. These are the structures produced due to erosion. The complexity of outcrops due to unconformity increases the potentiality for the formation of structural inliers and outliers.

EXERCISES

1 Work out maps No. 7, 8, 9, 10, 16, and 17; and draw suitable structure sections across the areas to illustrate the geological structure ?

2 Draw a section in the East and West line in map No. 2. Calculate the thickness of the beds A & B and indicate the inlier in the section.

3 Draw a structure section in map No. 4 along the line drawn in the direction NE—SW. Explain the behaviour of the course of outcrops of strata under the influence of their gradient and topography of the area ?

4 In map No. 3, draw a section along the line from SW—NE corner. Calculate the thickness of the beds and explain the geological structure of the area ?

7

FAULTS

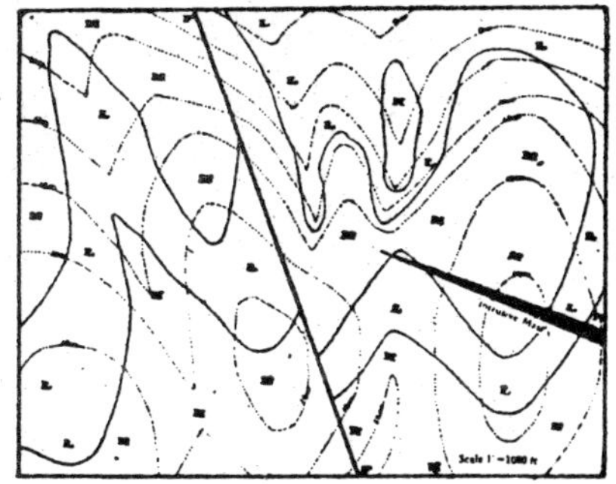

Fig : 74 Showing a fault running NW-SE.

Definition and Geometry of a Fault.

In the course of Jigantic earthmovements, strain may pile up till it exceeds the elasticity of rocks. Under such circumstances, fractures may be developed along which there has been relative displacement of the rocks. **Faults are therefore, fractures which displace the rocks.** The beds on one side of a fault may be Vertically displaced hundreds or even thousands of feet relative to the beds on the other side. The displacement in certain cases is horizontal and covers a distance of many thousand feet. The plane along which displacement takes place is known as a **fault plane.** The angle which a fault plane makes with a Vertical plane is its **hade** while that it makes with a horizontal plane is its dip. The angle of dip and angle of hade are compliment of each other. The amount of displacement of formerly adjacent points measured along the fault surface is called the **slip*** of the fault. The vertical component of separation between the upthrow and the downthrow sides along the dip of the fault plane is called the **throw** of the fault, while the horizontal component of dip separation between the upthrow and the downthrow sides is known as **heave.** The heave and throw are measured on a vertical section perpendicular to the strike of the fault plane. A vertical fault has no heave and its throw will be the same as the slip. The **offset** is the horizontal displacement at right angles to the strike of the beds.

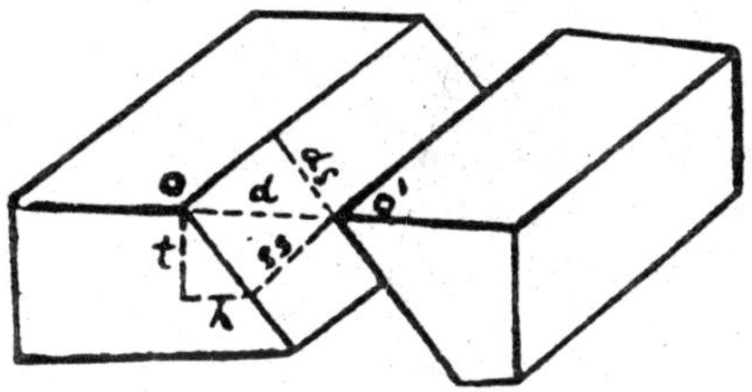

Fig : 75 Illustrating the geometry of a fault.

d=Displacement
ds=Dip Slip
ss=Strike slip
h=Heave
t=Throw

Descriptive Classification of faults

Normal & Reverse Fault—If the dip of the fault is in the direction of the downthrow, it is a **normal fault.** (Fig : 76) On the contrary, if the dip of the fault is in the direction of the upthrow, it is a **reverse fault.** A reverse fault, with a low angle of dip is known as **thrust.** The thrust is of two types i. e. **Overthrust** and **Underthrust.** In the former, beds apparently thrust over the other rocks, while in the latter case, the beds thrust under the other rocks

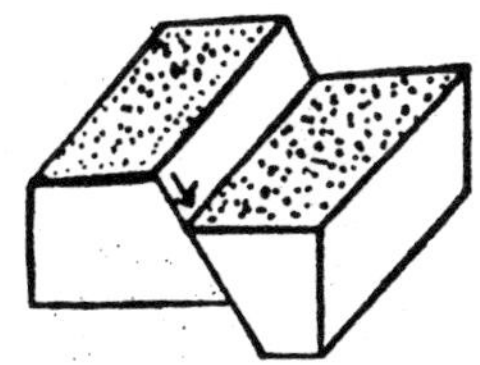

Fig : 76 Showing a Normal Fault

*The slip is measured on the fault plane and termed accordingly as Net slip, Dip slip, Strike slip, Vertical slip, horizontal slip, horizontal dip slip etc.

Dip Fault—The fault plane strikes parallel to the direction of dip of formations.

Strike Fault—The fault plane trends parallel to the strike of the formations.

Oblique Fault—The trend of the fault plane is Oblique to the strike of the formations.

Dip slip Fault—The relative movement of the fault blocks is along the dip direction of the fault plane.

Strike slip Fault—The movement induced due to faulting is parallel to the strike of the fault plane, It is also Called a Wrench or Tear fault.

Oblique slip Fault—The movement of the fault blocks is oblique to the strike and dip of the fault planes.

Step Fault—Several normal faults hading in the same direction give rise to a **step fault.** If there be two series of step faults which converge, the resulting structure is known as **graben** (or **trench** or **trough**). On the other hand, if there be two series of step faults which diverge, the resulting structure is known as a **ridge fault** (or **horst**).

Block Fault—If a block is bounded on two or more sides by gravity faults the resulting structure is known as a **block fault**.

Rotary Fault—When faulting is attended by a rotational movement, thus causing downthrow at one end and an upthrow on the other end of the strata; it is called a **rotary fault** (or **Scissor** or **pivotal** fault).

Hinge Fault—It shows displacement at one end, while the other end remains at its original place. The throw is nil at the hinge and goes on increasing along the fault line farther from the hinge.(see Fig : 77).

Fig : 77 Showing a hinge fault.

Inclined & Vertical Fault— If the fault plane is inclined, it is known as **inclined fault**. On the contrary, if the fault plane is vertical, it is known

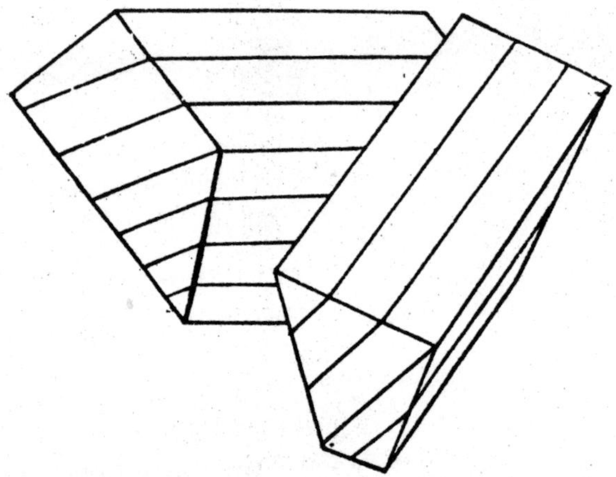

Fig : 78 Showing a Rotary fault.

as vertical fault.

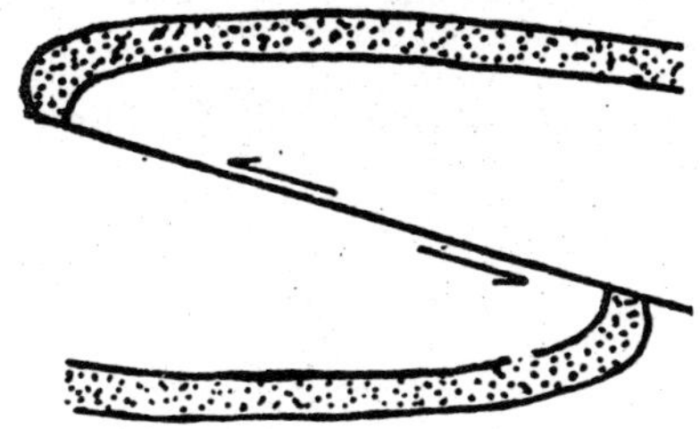

Fig : 79 Showing thrusting of beds.

Nappe, Klippe & Window—Due to overthrusting rock sheets are brought forward, which are known as **nappe.** Isolated remnants of the old nappe, which have escaped prolonged erosion, are designated as **Klippe.** In certain cases, due to erosion, rocks below a nappe may be exposed and such an eroded rock gap is known as a **window**.

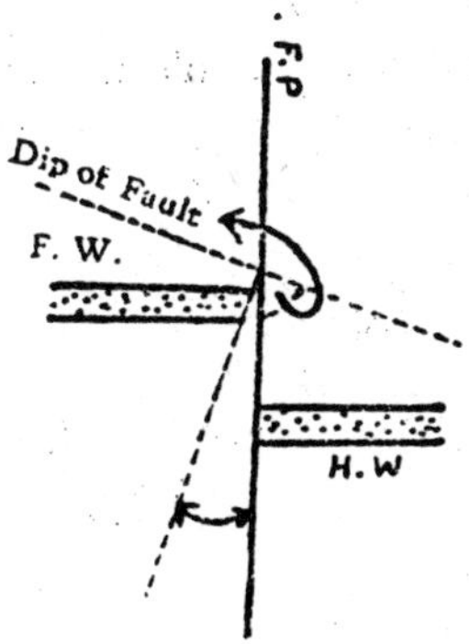

Fig : 80 Showing hanging wall and Foot wall.

Hanging Wall & Foot Wall

The block above a fault plane is known as the **hanging wall**, while the block below a fault is the **foot wall.**

Upthrow & Downthrow sides of a Fault—In a fault, the side that relatively goes upward is called **upthrow** while the side that goes down is called **downthrow.**

How to recognise the Upthrow and downthrow sides on a map—The side, along the fault plane, where older beds occur is the upthrow side, while the side in which the younger beds occur is the downthrow side This law fails, where the dip of fault is less than the apparent dip direction of the fault.

Calculation of throw - Draw strike lines for a bedding surface on one side and also draw strike lines for the same surface on the other side of the fault The difference between the two would give the amount of the throw.

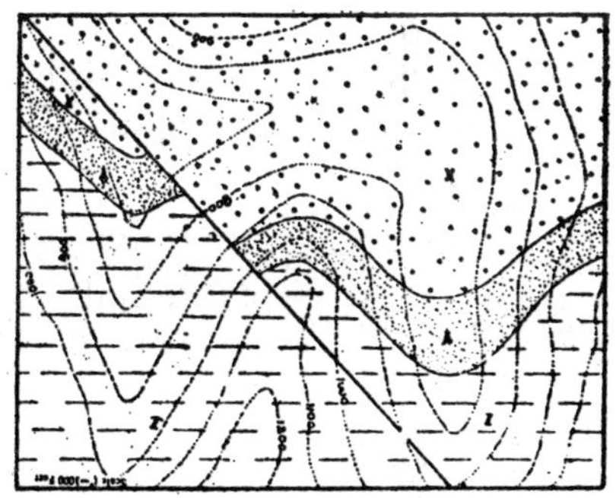

Fig : 81 Illustrating displacement where the throw is about 500 feet.

Effects of Faults On the Outcrops :

(i) Displacement of strata ultimately causes displacement of outcrops.

(ii) There is shifting in the position of the beds on either side of the fault.

(iii) Outcrops terminate abruptly.

(iv) Omission and Repetition of outcrops.

(v) Due to erosion, the younger beds are removed from the upthrow side and are preserved on the downthrow side of the fault.

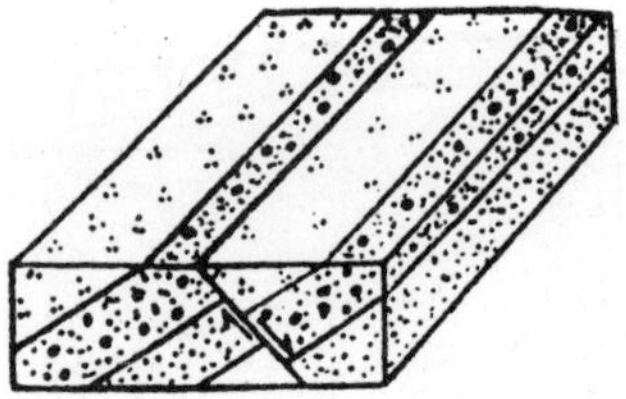

Fig : 82 Showing Repetition of beds.

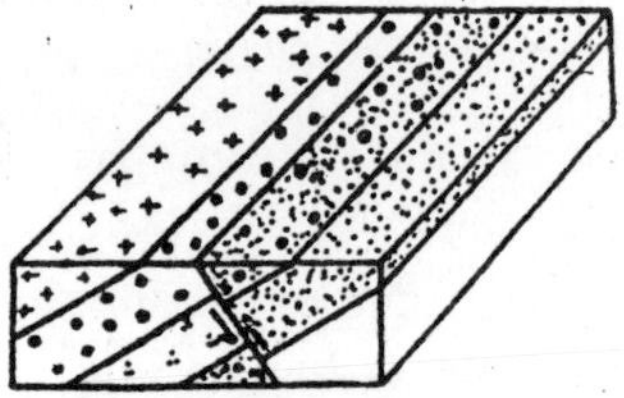

Fig : 83 Showing Omission of beds

(vi) Outcrops in synclines are pushed farther apart, while in anticlines, they are brought closer on the downthrow side. Opposite effects are noticed, if the slope of the surface is higher than the dip.

(vii) When slope of the surface and the dip of the beds are in the same direction, but dip is higher than the slope, the outcrops of the beds on the downthrow side appear to shift against the direction of their dip.

(viii) When the slope of the surface and the dip of the beds are just opposite in their directions, the outcrops of beds on the downthrow side appear to shift against the direction of their dip.

(ix) When the slope of the surface and dip of the beds are in the same direction, but the slope is much higher than the dip, the cutcrops of the beds appear to shift in the direction of dip.

(x) There is often change in the **sedimentary facies** as also in the **grade of metmorphics.**

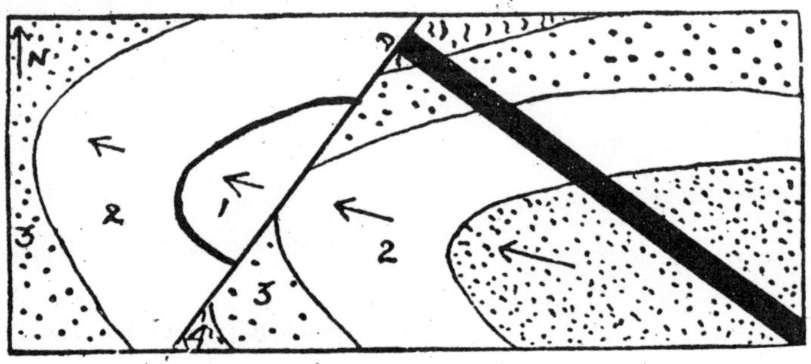

Fig : 84 Illustrating a fault on a map. The dips of beds being shown by arrows. The faulted beds are also intruded by a dyke. The upthrow side is to the west.

(xi) On the downthrow side of a fault, the outcrops of a younger bed terminates against the older bed in the corresponding upthrow side.

Attitude of the Fault—In faulted areas, the attitude of the beds **differ from** place to place. The amount of displacement in the direction of dip and strike varies. There are truncations of structural **features by the younger rocks.** There are enechelon zones of folds and fractures. In unconformable strata, the younger beds truncate older beds, whereas in areas of fault, the older formations truncate the younger beds. The outcrop of a faulted plane, as represented on a map, is a thick straight line.

Age of the Fault—A fault is always younger in age than the youngest beds it cuts. If the beds are superposed on a fault, concealing the later and remaining uneffected, are younger than the fault. Fault are **posthumous**, if further movement takes place along an existing fault plane. When faults cross each other, the one which is offset in the older.

Review Questions—

1. Draw structure sections of maps 11, 12, 13 & 14?

8

Folding

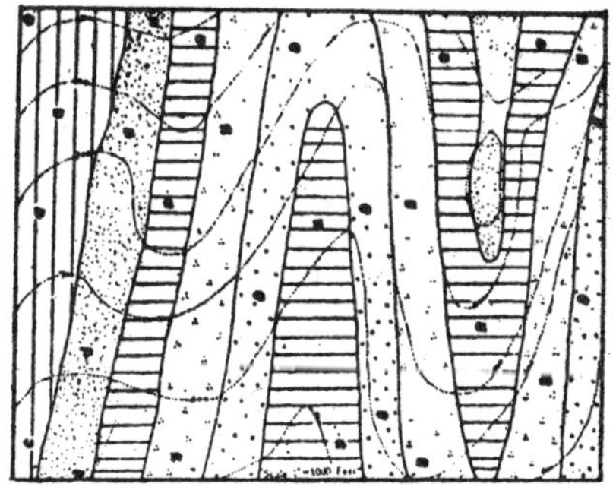

Fig : 85 Showing a map of folded beds.

Definition & Components of a Fold—A fold is an undulation in the rocks, produced due to lateral pressure brought about during the earth movements. Strata are often folded along an **axis of folding**. The axis of the fold may be tilted or horizontal, or pitch in one or two directions The plane that divides a fold is known as an **axial plane**, which may be Vertical, horizontal or dipping. The line joining the lowest points of a fold is known as the **trough line**, while the line joining the highest points of a fold is known as the **crest** line. The **flank** is a term applied to the side of a fold.

Descriptive Classification of Folds

Anticline & Syncline—When the beds are bent upward into an arch shaped structure, it is called an **anticline**. When the beds dip inward towards each other, the structure is known as **syncline**.

Symmetrical and Asymmetrical Fold—When the axial plane is vertical, the fold is upright and **symmetrical**. On the other hand, if the dips of the flanks of a syncline are unequal and the axial plane is inclined, the fold is **asymmetrical**.

Open and Close Fold —An **open fold** is one which may be bent closer without introducing flowage. On the contrary, a **close fold** is one in which the beds can not be bent closer without introducing flowage.

Recumbent Fold—If the axial plane assumes more or less horizontal position, the fold is said to be **recumbent**.

Isoclinal Fold and Overturned Fold—An **isoclinal fold** is one in which the two limbs are essentially parallel. The axial surface may be inclined or upright. If two limbs dip in the same direction at unequal angles and the axial plane is inclined, it is said to be an **overturned (or Over) fold**.

Homocline and Monocline Fold—A **homocline fold** offers a group of beds which dip in one and the same direction. It never represents a complete structure, but a limb of either anticline or syncline. The **monocline fold**, on the other hand is a step like bend in otherwise horizontal beds.

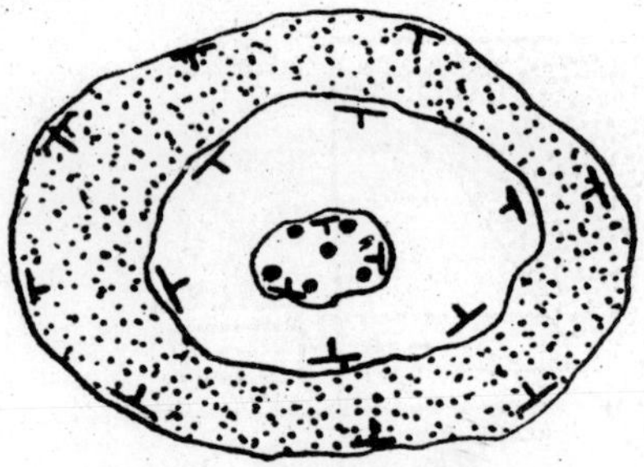

Fig : 86 Showing a Basin.

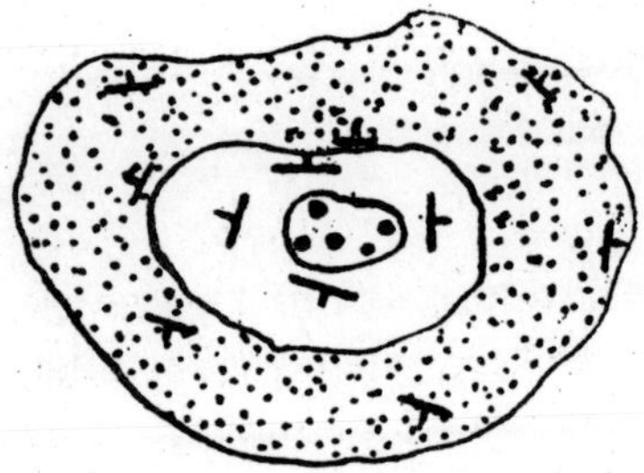

Fig : 87 Showing a Dome.

Dome & Basin –

When beds buckle or bend in such a pattern that the dip is everywhere outward from a central point, the resulting structure is known as a **dome.** On the contrary, when dip is every-where inwards from a central point, the resulting structure is known as a **basin.**

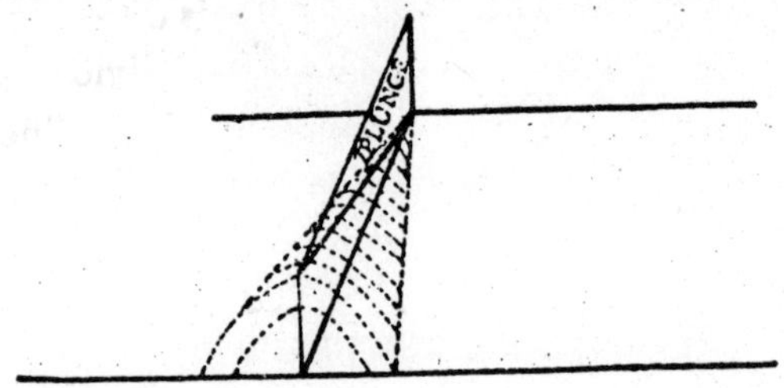

Fig : 88 Illustrating plunge or pitch of the fold.

Plunging Flod—These are quite limited in their range and may plunge at either end before termination. If the structure is an eroded anticline, the **pitch** or **plunge** is in the direction of the convergence of the limbs. In an eroded syncline, the plunge is in the direction of the divergence of the limbs. The plunge can be found out by comparing the attitudes of two or more points on the outcrop of a bed, where it crosses the fold axis. The calculation of the plunge can be made from the spacing of the strike lines, measured in the direction of the plunge.

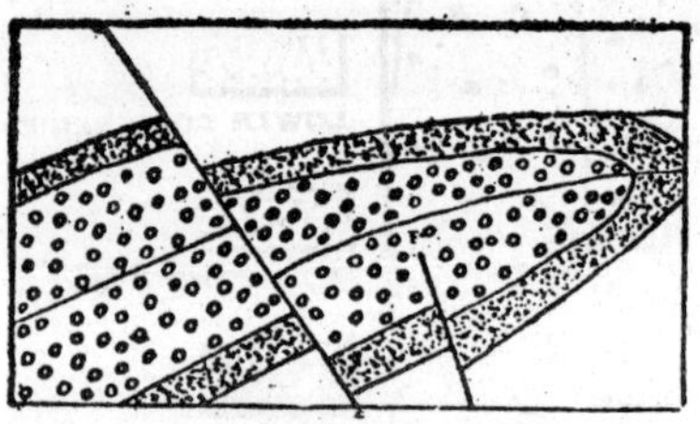

Fig : 89 Illustrating the outcrops of an anticlinally folded beds, plunging to the East.

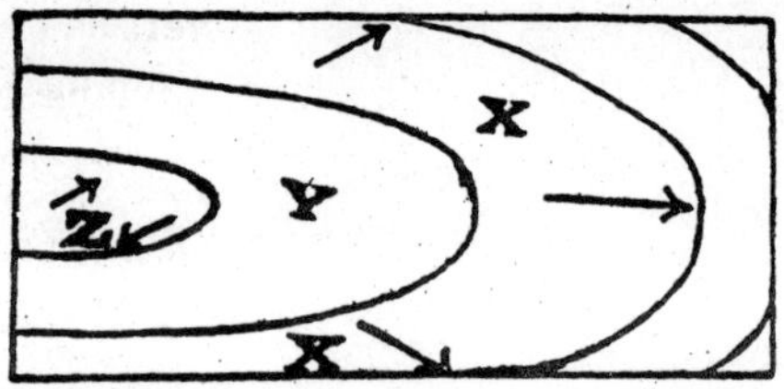

Fig : 90 Showing three beds Z, Y & X plunging towards the convergence of the limbs.

Outcrops of Folded Strata

Folded strata have their outcrops repeated and dips of the two outcrops will be in the opposite directions. In an anticline the older beds will outcrop in the centre, while in a syncline the youngest beds will outcrop in the centre. The basin and dome have elliptical outcrop. Folded strata have two sets of stratum contours i.e., one for each flank of the fold. In several cases, the dip of the strata changes in amount gradually as the axial plane is approached. In such cases, the structure contours would not be equi-spaced.

(By Courtesy of M/s George Allen & Unwin Ltd London, From the Book 'Elementary Exercises upon Geological Maps" by John I-Platt).

Fig : 91 Showing formations with folded structure.

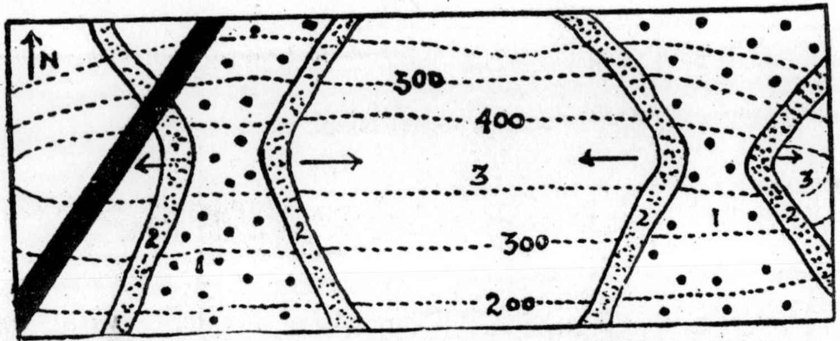

(By courtesy World Press Private Ltd; Calcutta)

Fig : 92 Showing the folling of beds. The beds 1--3 are arranged from older to younger. The outcrops of bed 2 show opposite directions of dip in the successive outcrops. The folded area is also intruded by a dyke.

The folds with horizontal axes are subject to weathering and the outcrops of the beds on two sides of an axis are almost parallel. On the other hand, if the axes are inclined, as in the pitching folds, the two outcrops of the bedding surfaces of a stratum on a nearly flat surface converge towards each other, in the direction of the pitch.

Due to folding, dip directions of the outcrops Varies. In an anticline, two limbs dip away, while in a syncline, two limbs dip towards, the axial plane.

EXERCISES

1. In map No. 14 all structural features such as fold, fault, unconformity etc; are present. Draw a structure section and write its history ?

2. Work out map No. 2, 3, 4, 5 and 7 and draw their sections ?

3. In Fig 91 draw a structure section and write the Geological history clearly explaining the structure of the area ?

9

Igneous and Metamorphic Rocks

Igneous rocks may be defined as rocks which have solidified, with cooling from magma or lava, with or without actual or apparent crystallisation.

According to their mode of occurrence, they are classified into two groups—INTRUSIVE and EXTRUSIVE. The intrusive rocks include the plutonic and hypabyssal, while the extrusive rocks include volcanics. The intrusives may be **concordant or discordant.** The concordant igneous bodies include sill, laccolith, lopolith, phacolith, etc , while the discordant igneous bodies include **dyke, batholith and stock.**

Sill—A thin tabular sheet of magma, which penetrates along the bedding plane. It may range in thickness from a fraction of foot to several feet. The intrusion of a sill seldom causes disturbance of the strata, if at all there is slight tilting, so the sill is also tilted with the same amount of dip as that of a bed Sill is constituted of either doleritic or basaltic matter. On a map, the outcrop is parallel to the succession. A sill is younger in age than the rocks it cuts.

Laccolith—It is a lenticular body with an arched up roof and horizontal floor. On a map, it appears as a circular igneous body, younger than the overlying strata. It shows concordant boundary with the surrounding beds. The beds lying at the top and on the sides of a laccolith dip in all directions away from the igneous body.

Lopolith--It is a large flat topped concordant igneous body with sunken base. The rocks underlying and overlying a lopolith dip inward towards a common centre.

Phacolith—It is a concentric shaped concordant body, occurring along the crests and troughs of folds.

Dyke—It is a tabular shaped discordant igneous bodies. It cuts across the structures of the invaded strata. It may be vertical or tilted. On a map, it shows straight outcrop only at the time when the dyke is vertical otherwise curved outcrop. Dyke is younger than the rock it penetrates; but older than the rocks as are not affected by it. An arcuate dyke occurring in circle is called a **ring dyke**.

Sheets of igneous bodies, which occur in the form of inverted coaxial cones, with thin layers of rocks lying in between them, are known as **cone sheets.** These show some amount of dip towards the apex of the cone.

Batholith—It is a larger body of discordant type. It is exposed only after a prolonged weathering of overlying strata. The outcrop of a batholith is less regular and does not show any uniform dip. A batholith is younger than the rocks it penetrates.

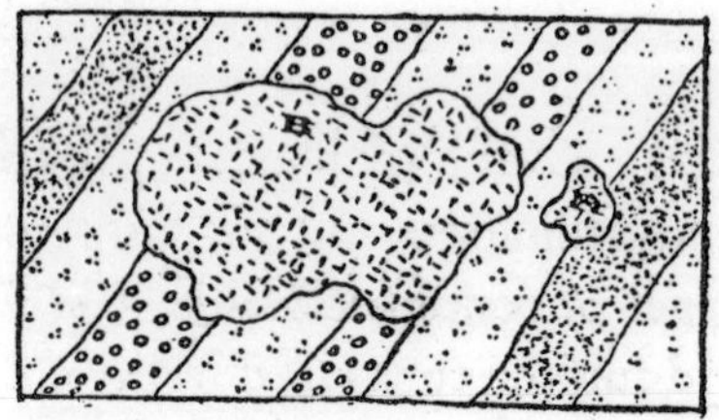

Fig : 93 Illustrating a Batholith on a map.

Boss—An intrusive mass of smaller dimension. It is plug shaped, with steep sides. It constitutes the core of hills. The outcrop is oval shaped.

Stock— An intrusive mass, injected deep into the earth's crust, It is visible only after a long erosion. On a geological map, its shape is quite irregular.

Lava Flows and Tuffs—A lava flow comes into being when magma ejects out at the earth's surface in a quiet manner. From a map alone, it is rather difficult to differentiate between a sill and a lava flow. The surface of the lava flow is no doubt smooth, but often covered by irregularities of different magnitude. When poured, on eroded surface, may consolidate and rest upon the beds of different periods, thus producing an unconformity. The lava flow is conformable and sill like, when poured on the sea bed. On the maps, it produces pattern similar to that of a sedimentary bed. **Tuff** is primarily a fragmentary matrix of ash and appears on a map, like a sedimentary bed despite its pyroclastic origin.

Volcanic Neck—It is a consolidated igneous mass, filling up the vents or pipes of volcanoes. It resembles a boss but is smaller in dimension. The rock types encountered are entirely different from that of a boss. In cross sections, it is almost circular and sides vertical.

Rocks around the intrusive and extrusive masses are often baked. The changes thus produced in the rocks are very remarkable. The baked rocks are also designated as the **metamorphic aureoles.**

EXERCISE :

Draw a structure section of map No. 14, along the line drawn in NW-SE direction. Also show the form of igneous rocks ? Find out the relative ages of the igneous rocks as far as it is possible ?

Interpretation of Geological Maps

As the geological map contains, in addition to the ordinary details of the topography, information regarding the nature, age,inclination, and general relationships of the rocks from the mantle, the interpretation, therefore, depends largely upon the types of contacts between rock units and how the appearance of these contacts varies as a result of differences in structure and surface topography. The sequence also plays an important role in the interpretation of a geological map. In reference to the attitude of the sequence, it is imperative to check up the position of the youngest and the oldest bed. The position, therefore, depends upon three factors.

First- If the beds are unconformable, the beds which appear to be trapped are the oldest, while those which trap are the youngest.

Second- If the older beds outcrop inside the younger ones, it is an **anticline.** On the contrary if, the younger beds outcrop inside the older ones, it is a **syncline**, and

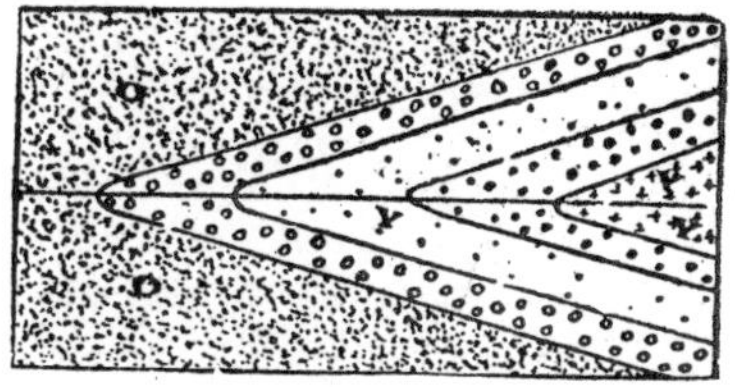

Fig : 94 Showing position of beds in a synclinal structure.

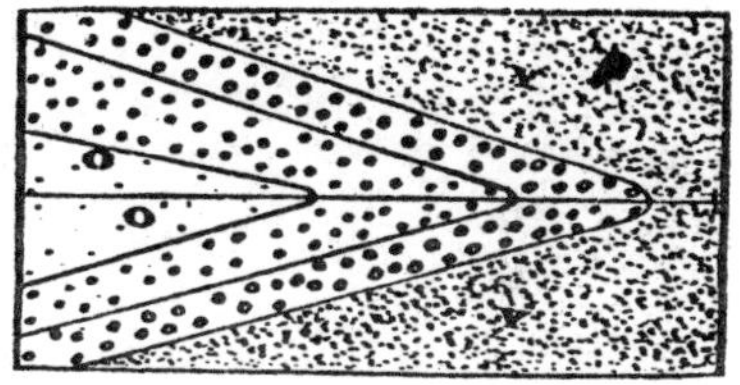

Fig : 95 Showing position of beds in an anticlinal structure.

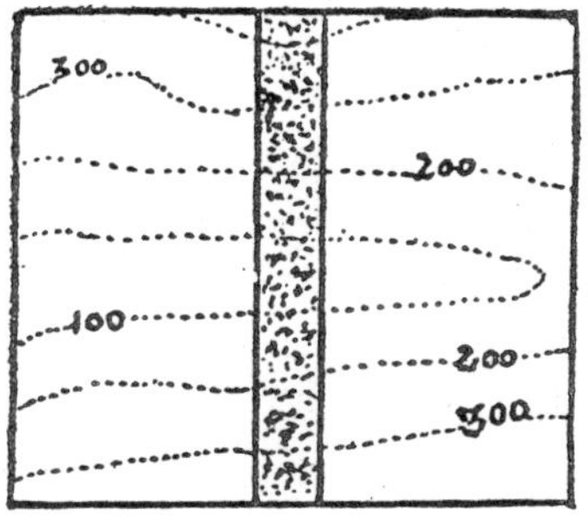

Fig : 96 Showing vertical bed.

Third- The tilted strata are youngest along dip, provided the dip is constant in a particular direction and dip is greater than the slope of the surface. If dip is lesser than the slope of the surface, the rocks are definitely older. The interpretation of a map, reveals that the beds are rarely horizontal but inclined.

A geological map contains a table of conventional signs which lists and explains the symbols on it. The exact form of the symbols used to represent the information varies, of course, in different surveys but the variation is limited and therefore, the general descriptions and principles serve with but minor modifications.

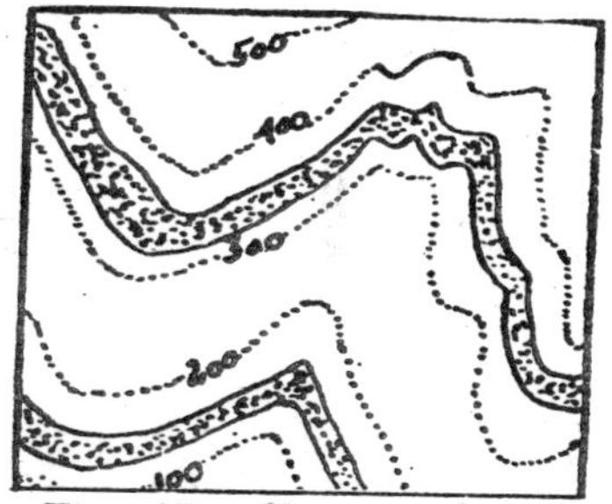

Fig : 97 Showing horizontal beds

1. A geological **BED** may be defined as a structural unit of sedimentary rock, having a small thickness, while the term **Formation** is applied for a bed having large dimension and also for a series or group of beds, in which all the beds have similar attitude but different from the attitude of any other group.

The maps containing different rocks, forming the surface of the earth, are shown by means of colour or by suitable shading. In addition, information is given by means of suitable symbols of the portion and extent of mineral veins, fractures, and unconformities, and of the altitude of the beds of stratified rocks with regard to the horizontal plains and their relations to each other. A colour scale to designate the age of rocks, proposed by Karpinsky, was adopted by the 1881 International Geological Congress. It gives different colours for the different stratum systems.

In order to interpret the map correctly and accurately, it should be examined carefully to understand the topography the influence of topography on the outcrops. Take the amount and direction of dip from many points, so as to study the variations and the positions of the axes of flexures, if there be any on the map. Try to work out, the presence of faults, igneous intrusions and their nature, so as to visulise the **solid geology** of the area. The horizontal sections, offer true picture of the underground structure. In drawing a horizontal section across the map the line of section should always be along the direction of true dip, the line of section should pass through the highest contours and also it should pass through as many beds as possible. So as to visulise the complete picture of the structure and topography of the area.

It is desirable that one should have an independent approach to the study and interpretation of the geological maps.

Construction of Map Section

Map section is completed in two phases. In the first phase, **profile is completed** and in the second phase, section is drawn. A profile is the vertical sectional view of the topography of an area along a definite line The profile is constructed by the help of four lines, namely **the base line, the profile line**, and the two end lines. The position of the profile section is marked on a map by a line known as the **line of section.** A strip of paper is laid along the line of section and the points where the section line cuts the contours are plotted and are then transferred to the paper on the given base line. Draw perpendiculars from the plotted points with respect to the height of the contours. The end points of the perpendiculars are then joined together with free hand. Make it a point to draw concave and convex curves for representing the valleys and the apex of the hills. Thus profile is complete.

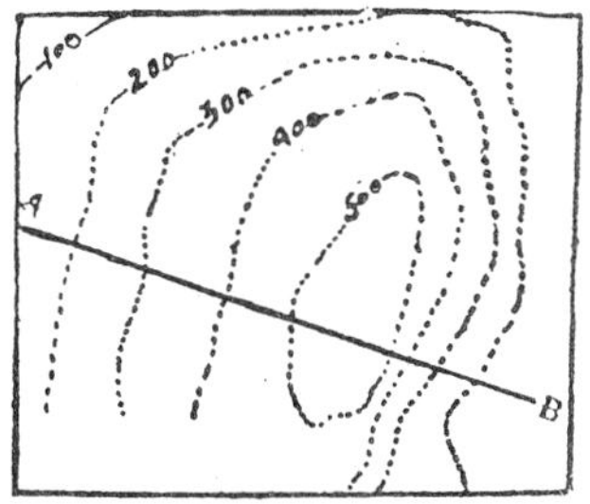

Fig : 98 Showing a sect-ion line A B

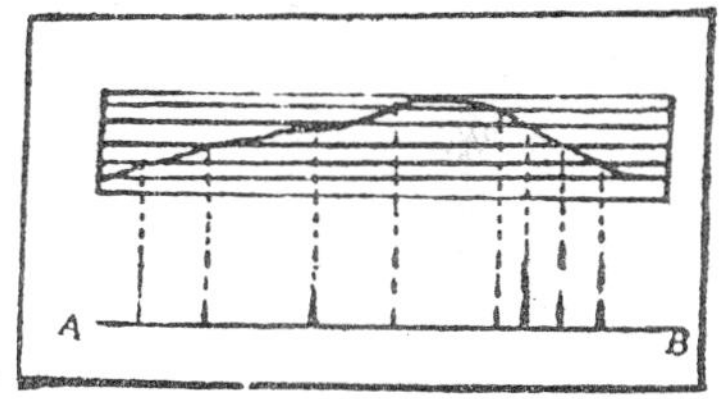

Fig : 99 Showing the construct-ion of a profile.

The section is drawn along a line of section, marked on the map. Here the points of intersection of the outcrops with the line of section are plotted and then transferred on the base line. Draw perpendiculars from these points and produce them so as to touch the profile line. Now draw lines, from the points (where the perpendiculars of outcrops touch the profile) with respect to their angles of dip and directions of dip. These lines thus represent the bedding surfaces of the strata, present on the map.

The scales used, the direction and position of the section line should also be marked at the margin of the structure section. Finally, symbols, signs, conventions and colours, as may be used on the map, should be used. In certain cases, the index or legend may not be given. it should be worked out with a great care, so as to represent the correct order of sequence on the section.

Thus a structure section showing the arrangement of the rock units in a vertical plane extending below the earth's surface is ready.

Geological History of the Map—Having drawn the section, give a concise account of the geological history of the area, for which map is drawn, under the following heads :

(i) **Sequence.** (ii) **Structure** (iii) **Geological History**

In sequence, work out, whether the strata are regular or irregular. (Conformable or Unconformable). Establish the relationship of the older and the younger strata. Find out unconformities and correct order of succession. Count the number of beds, their geographic directions and also the amount of their dips.

In structure, study the outcrop pattern, mention the type and trend of flexures, faults, outliers, inliers and other topographical and the structural features. Discuss the nature of intrusive and extrusive igneous bodies; which intersect the section line. Study and mention the specific type of a fold, its axis and direction of plunge. In case, map represents faults, mention the specific type, the direction of throw and also give the amount of the throw. Further refer to the attitude of the faults and their effect on the outcrops. Mention in history, the age of the fault. Explain in brief the relationship of the topography to the geological structures. If the close examination of the map, reveals the presence of rivers or streams then mention the specific type of drainage and direction of their flow. Mention map also be made, if their is any other special topographical or structural features.

In the history of the map, having given the position of the younger and the older beds, give the mode of formation of the beds. Explain in brief the trend of environments in which the sedimentation has taken place. If the area is subject to diastrophic or crustal movements or if there is an emergence or submergence of land, then trace out their original history and discuss in detail, the existing configuration of the relief. Special points, if any, may also be explained in the geological history.

EXERCISES UPON GEOLOGICAL MAPS

ELEMENTS OF GEOLOGICAL MAPS NO. 1

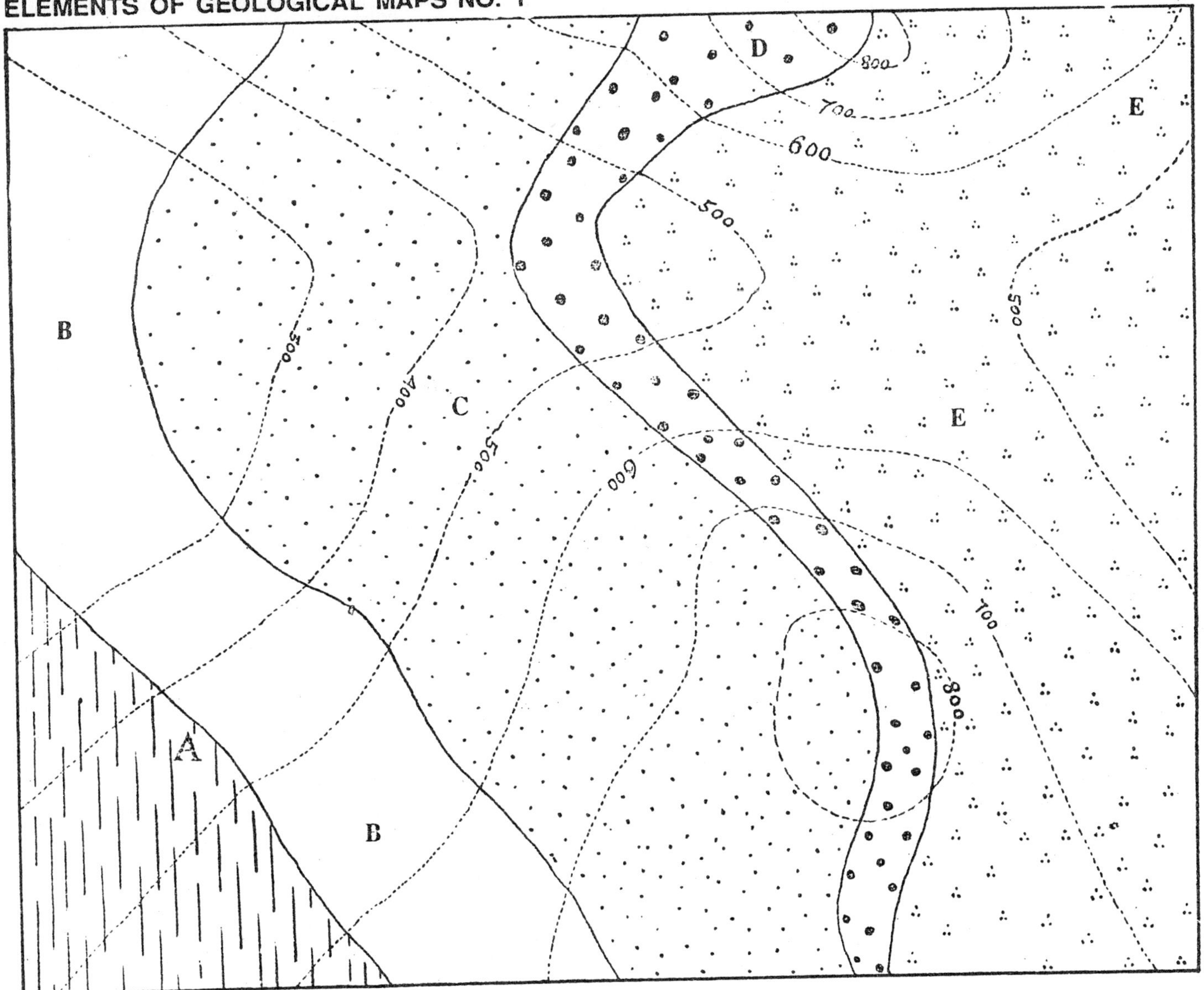

ELEMENTS OF GEOLOGICAL MAPS NO. 2

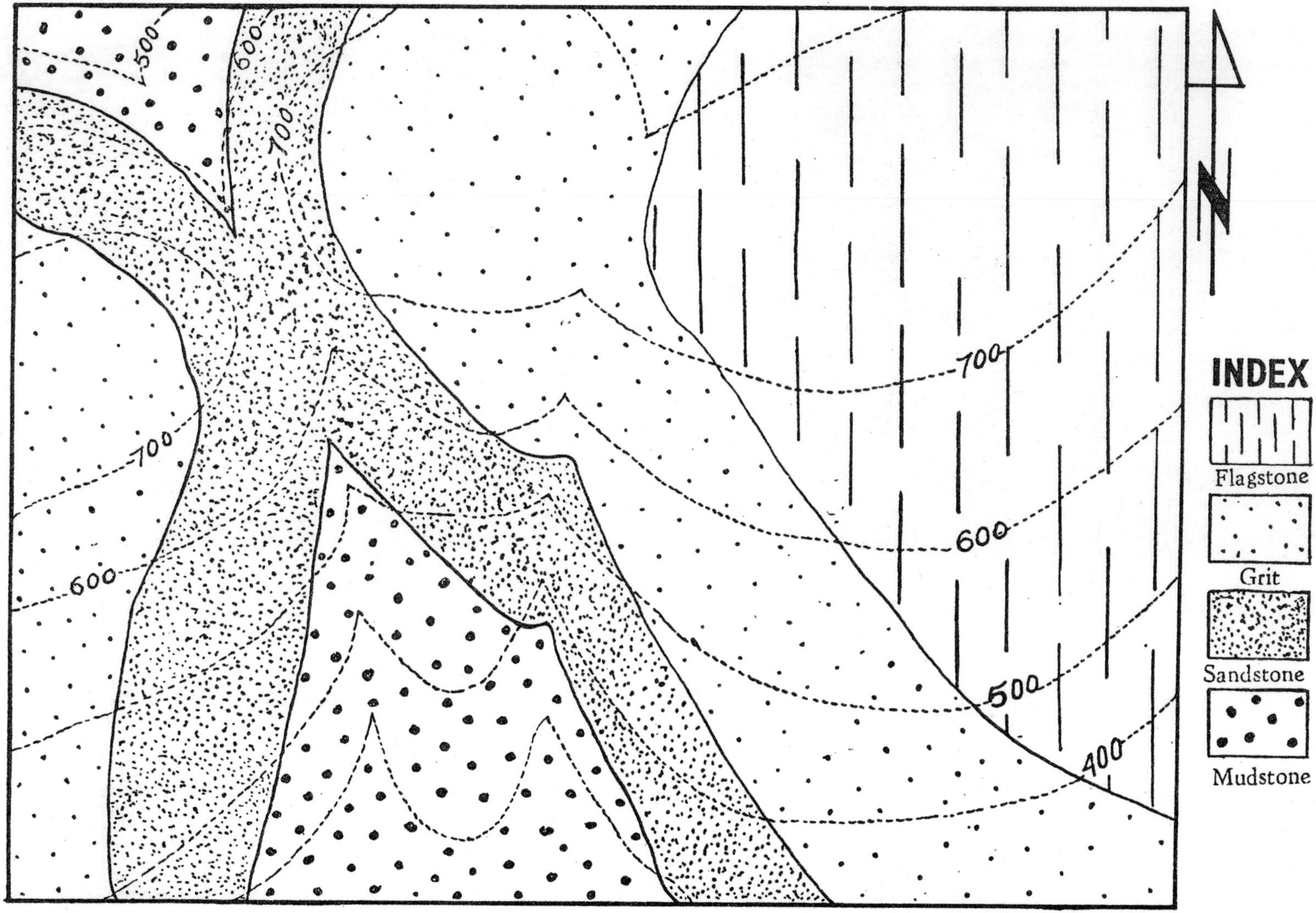

By courtesy of Prof. T. N. Muthuswami.

Scale 1 inch : 1000 feet.

ELEMENTS OF GEOLOGICAL MAPS NO. 3

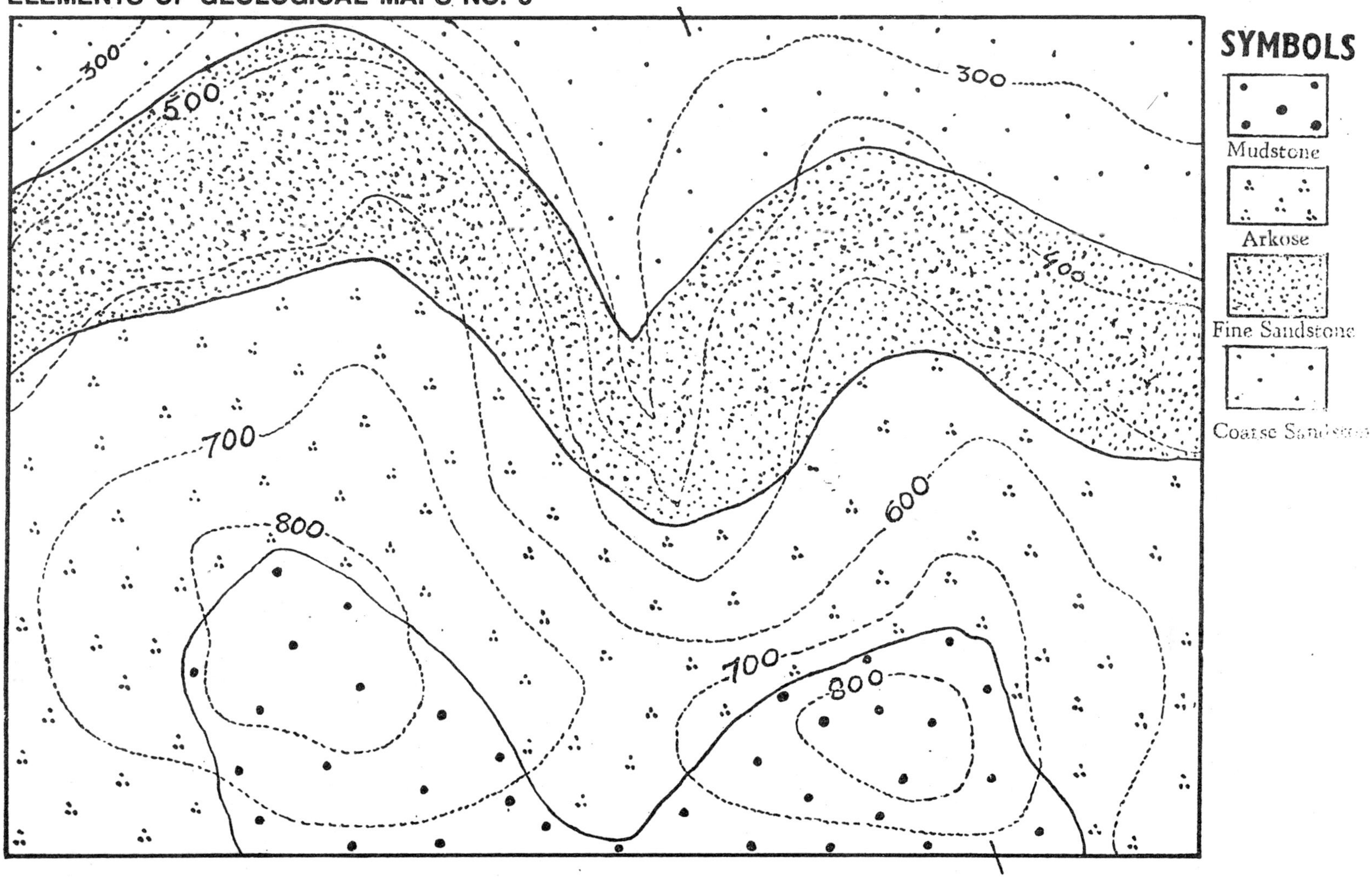

ELEMENTS OF GEOLOGICAL MAPS NO. 4

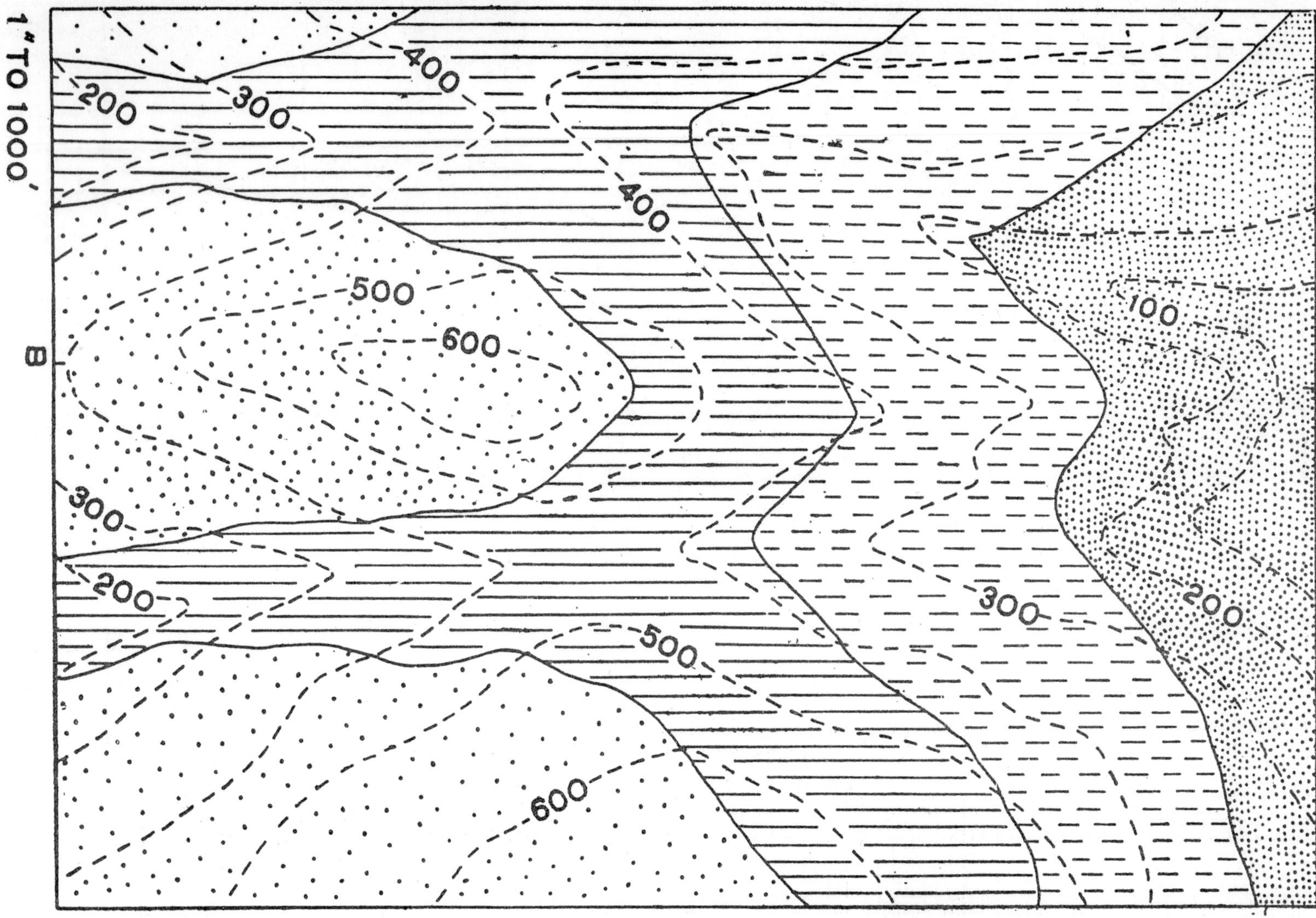

ELEMENTS OF GEOLOGICAL MAPS NO. 5

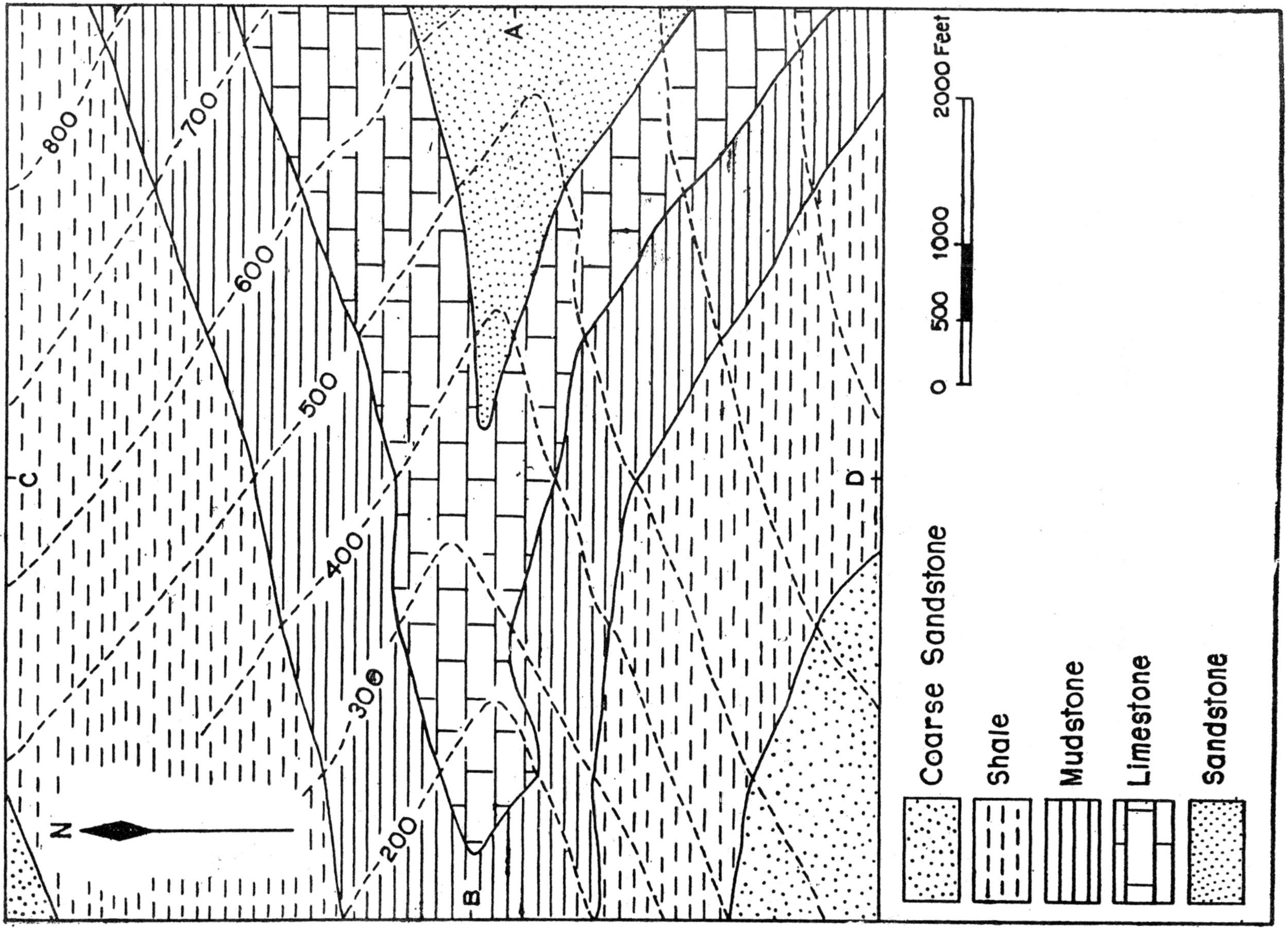

ELEMENTS OF GEOLOGICAL MAPS NO. 6

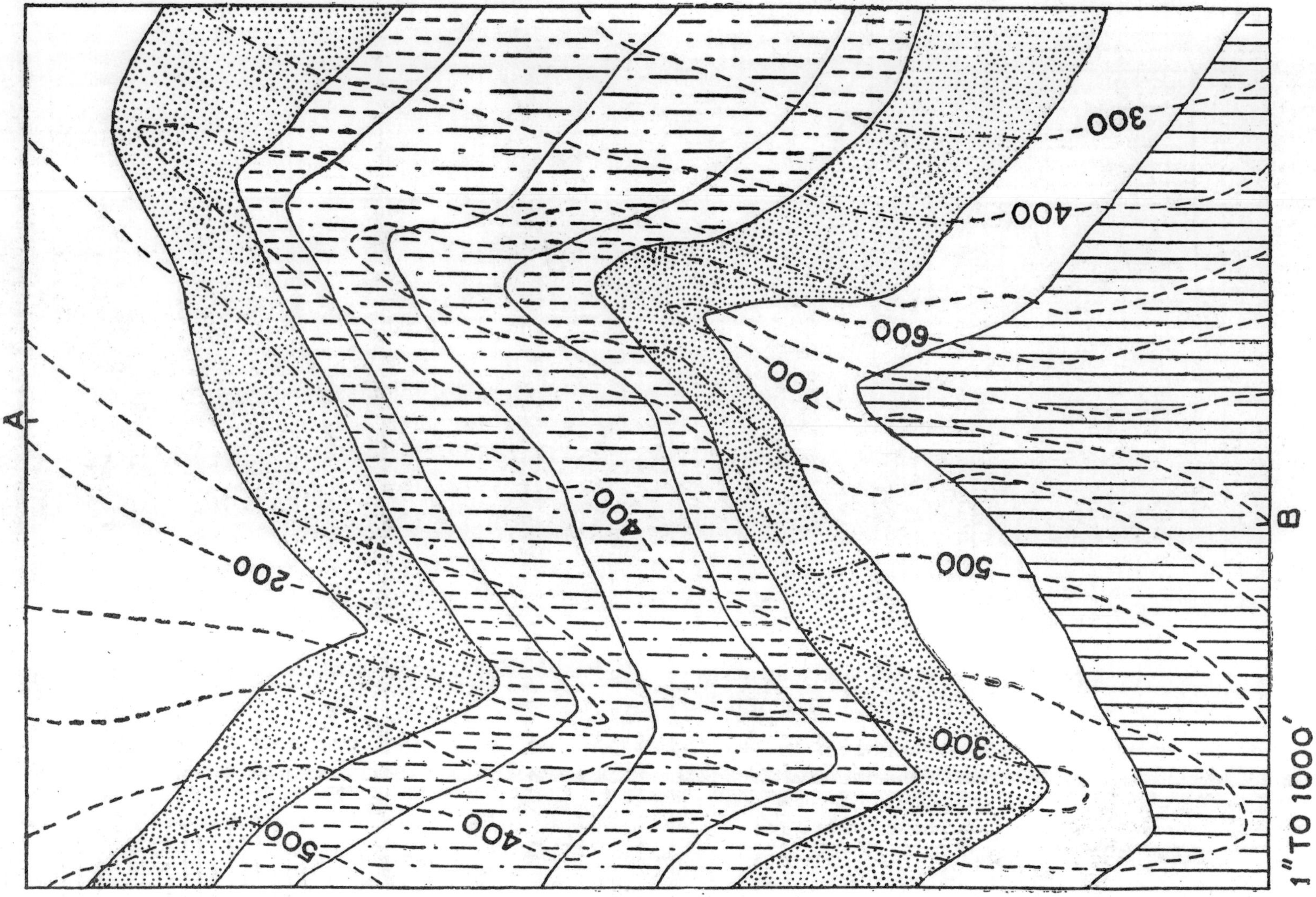

ELEMENTS OF GEOLOGICAL MAPS NO. 7

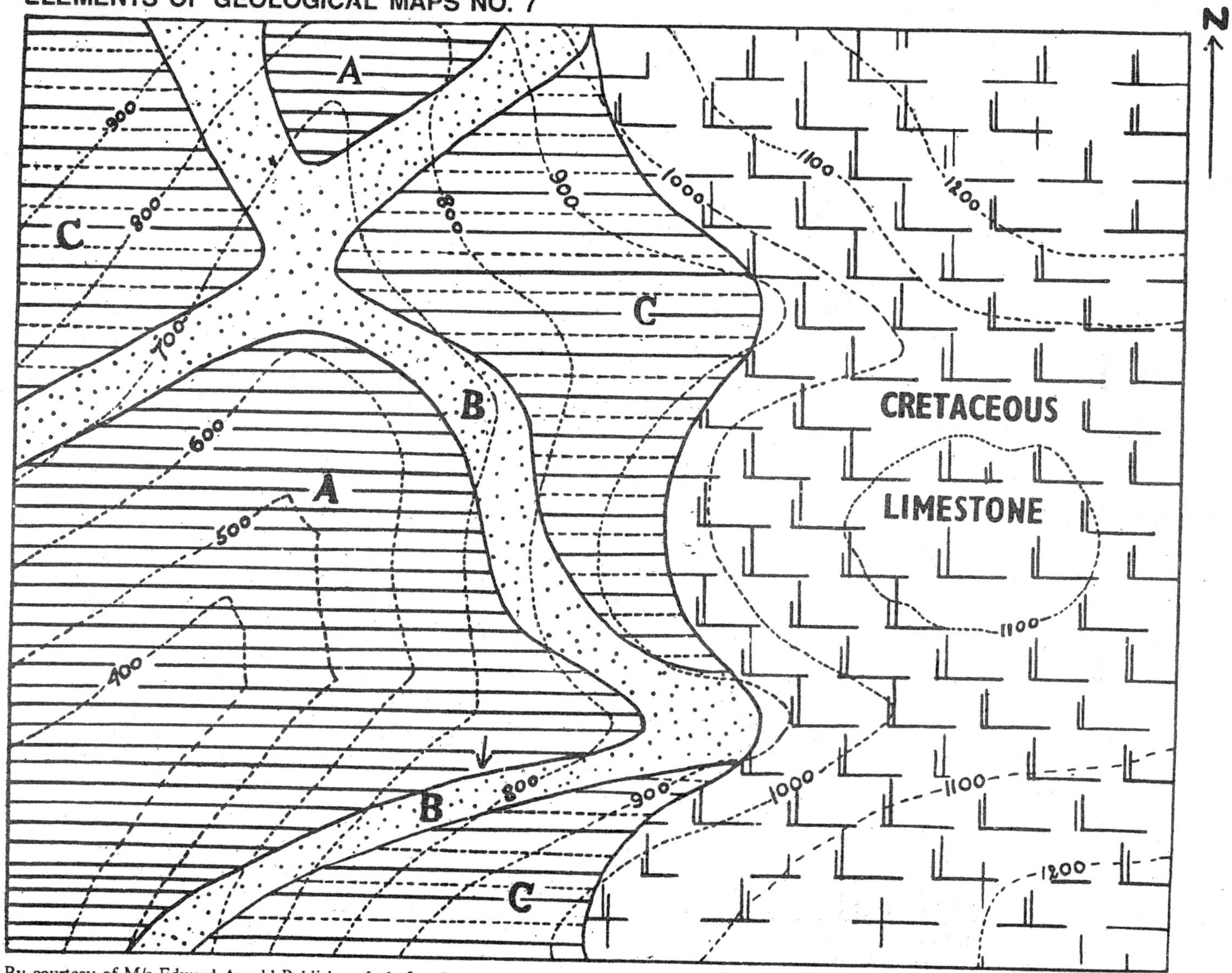

By courtesy of M/s Edward Arnold Publishers Ltd., London
(From the book "An Introduction to Geological Structures & Maps" by G.M. Bennison).

ELEMENTS OF GEOLOGICAL MAPS NO. 8

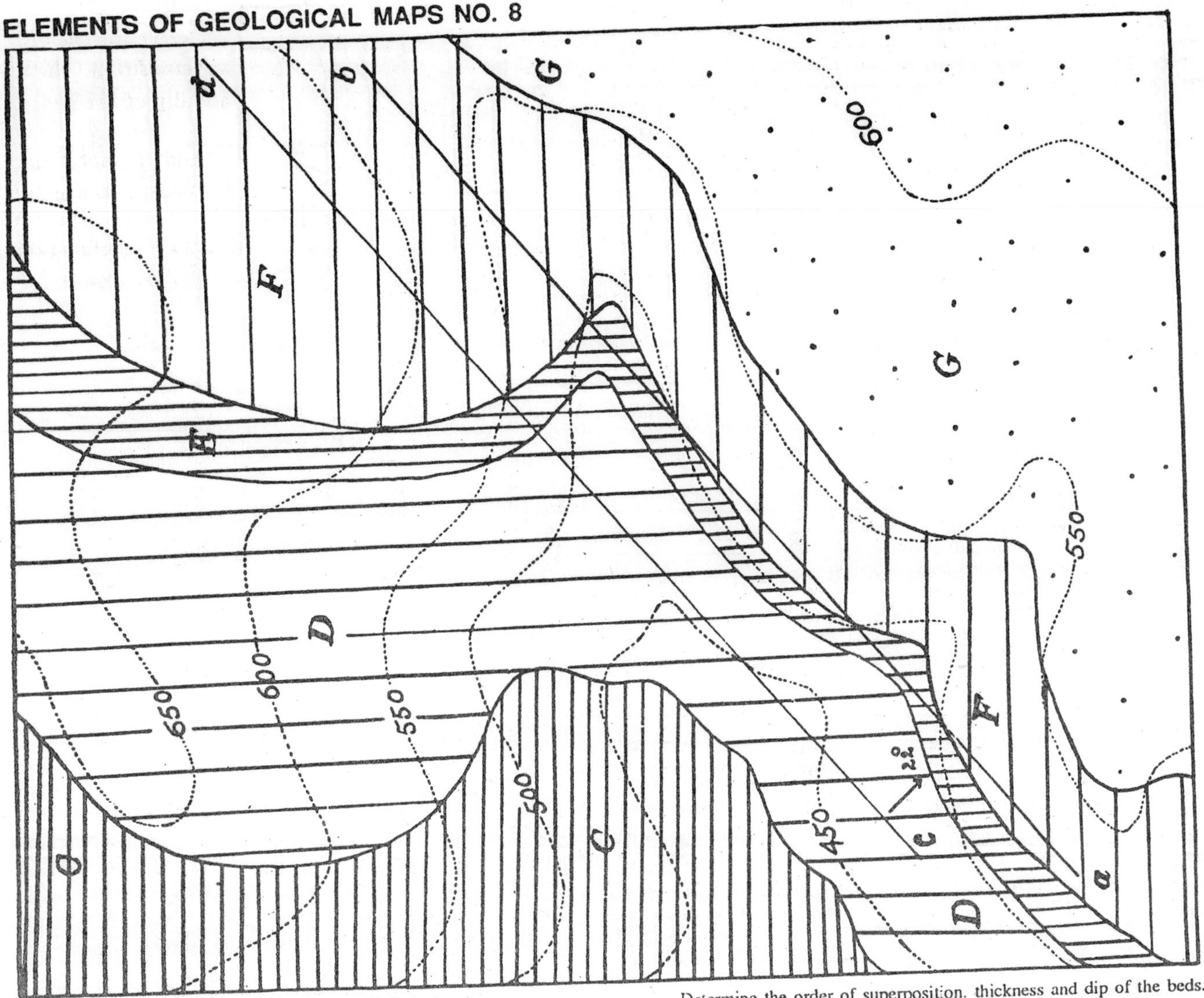

(By courtesy of M/s Edward Arnold Publishers Ltd., London
(From the book "A Geology for Engineers" by F.B.H. Blyth
— 1962 edition).

Determine the order of superposition. thickness and dip of the beds.
Add a note on the geology of the area.
Draw a suitable section. Scale 1" = 1000 ft.

ELEMENTS OF GEOLOGICAL MAPS NO. 9

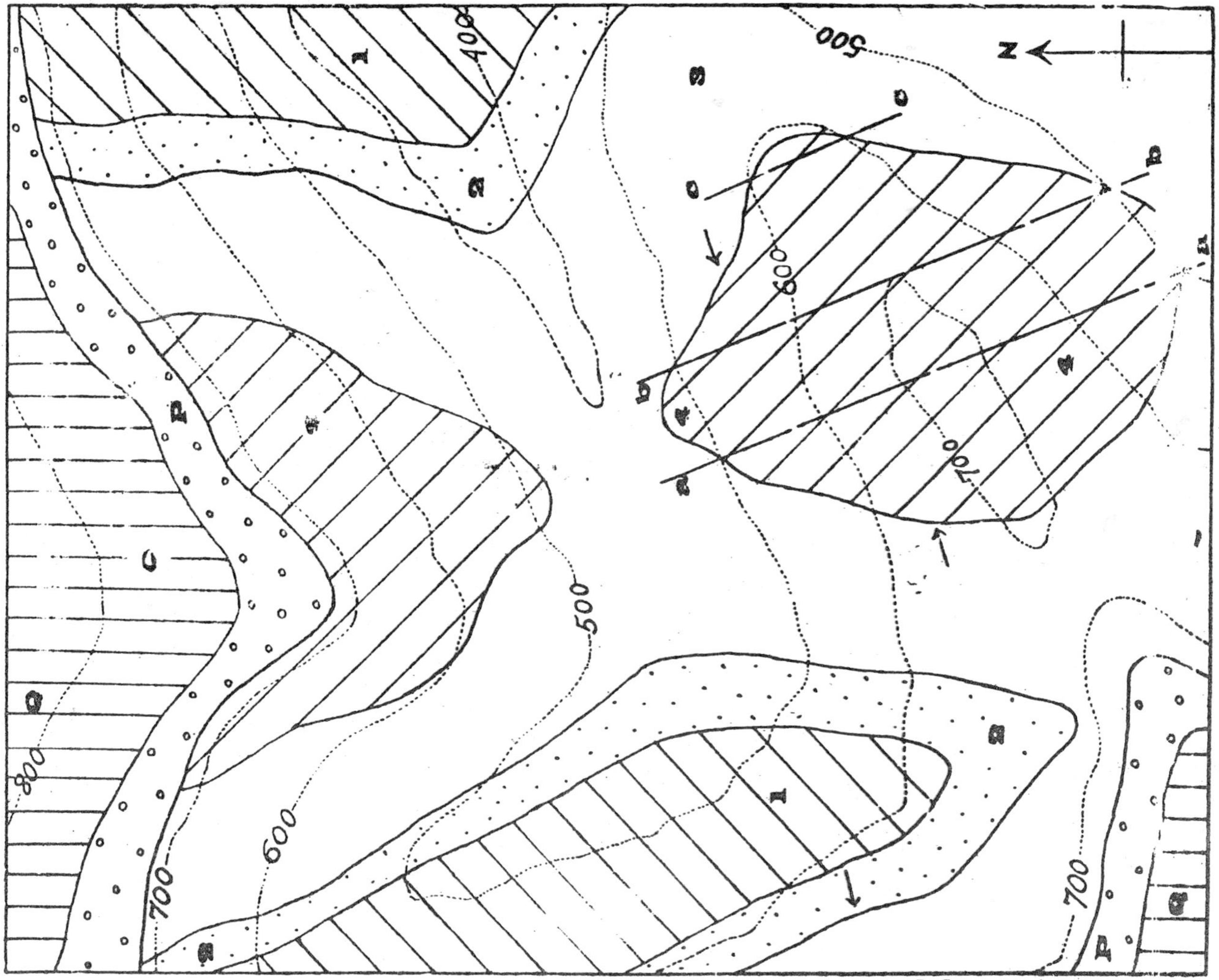

Determine the order of superposition, thickness, and dip of the beds.

Add a note on the Geology of the area.

Draw a suitable sections

Scale 1"=1000 ft.

By Courtesy of M/s EDWARD ARNOLD (Publishers) Ltd; London. (From the book "A Geology for Engineers" by F. G. H. Blyth-1962 Edition).

ELEMENTS OF GEOLOGICAL MAPS NO. 10

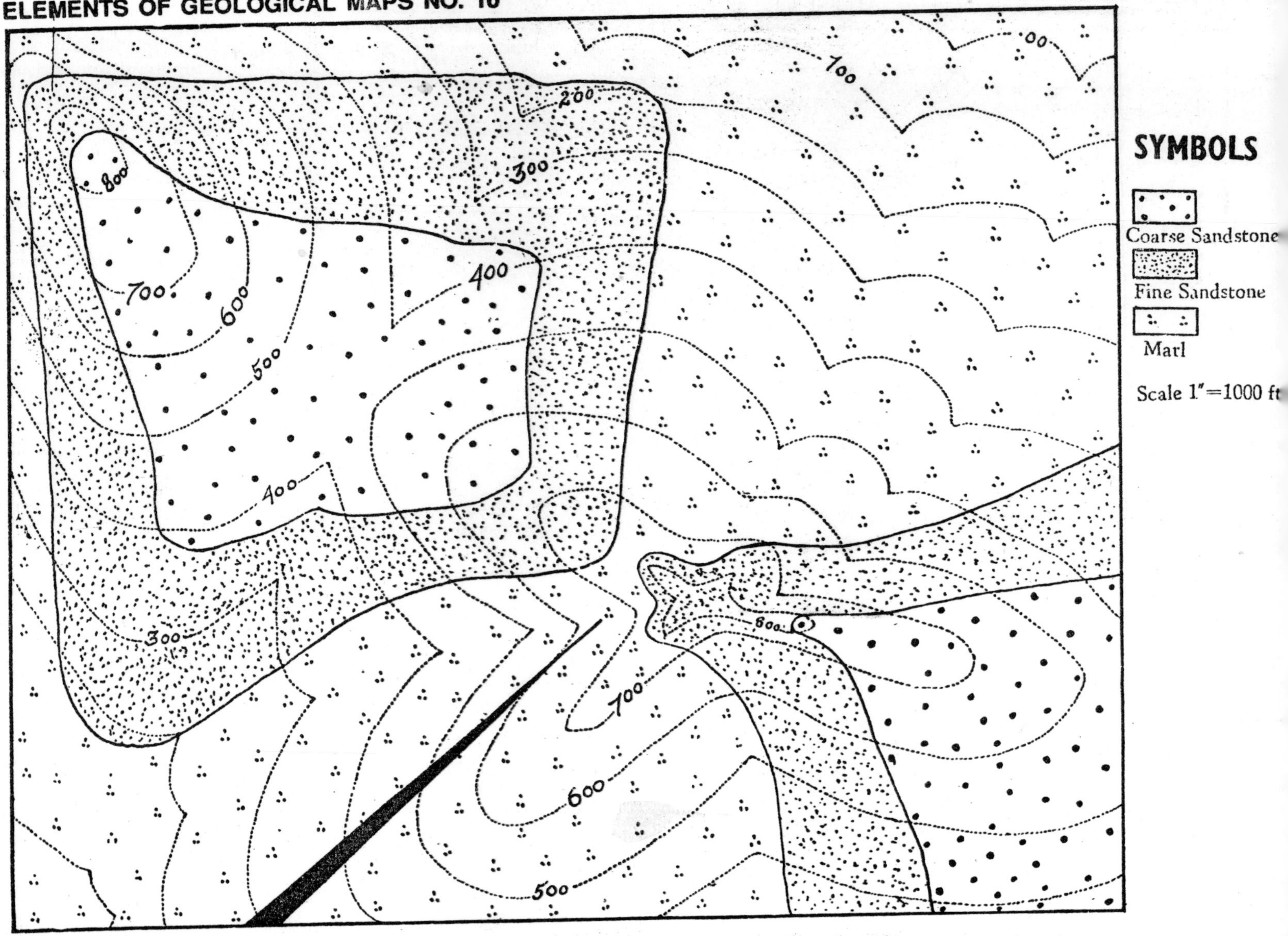

ELEMENTS OF GEOLOGICAL MAPS NO. 11

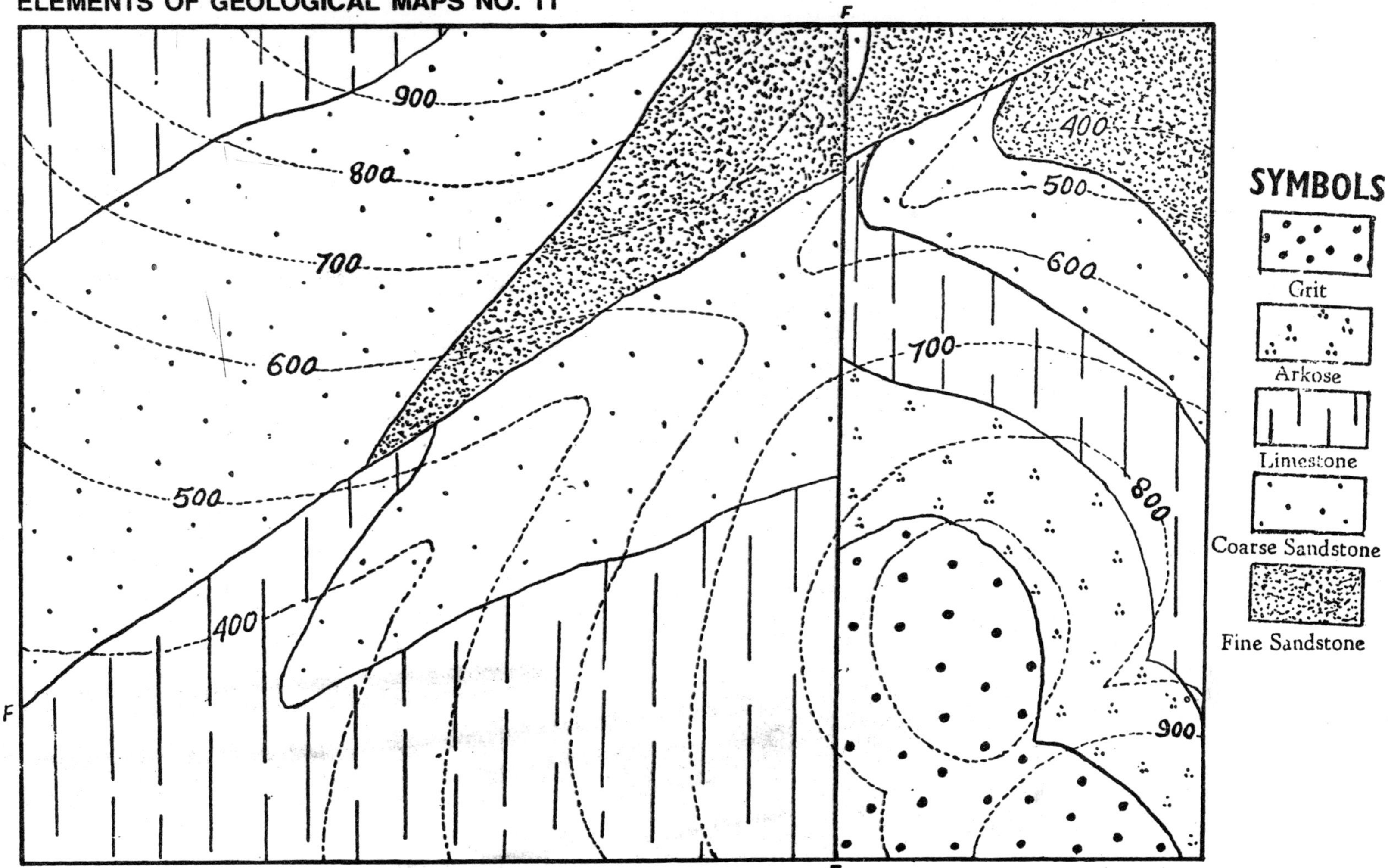

Describe the geology of the area giving nature & sequence of the two faults Draw a suitable section.

Scale 1 inch : 1000 feet.

ELEMENTS OF GEOLOGICAL MAPS NO. 12

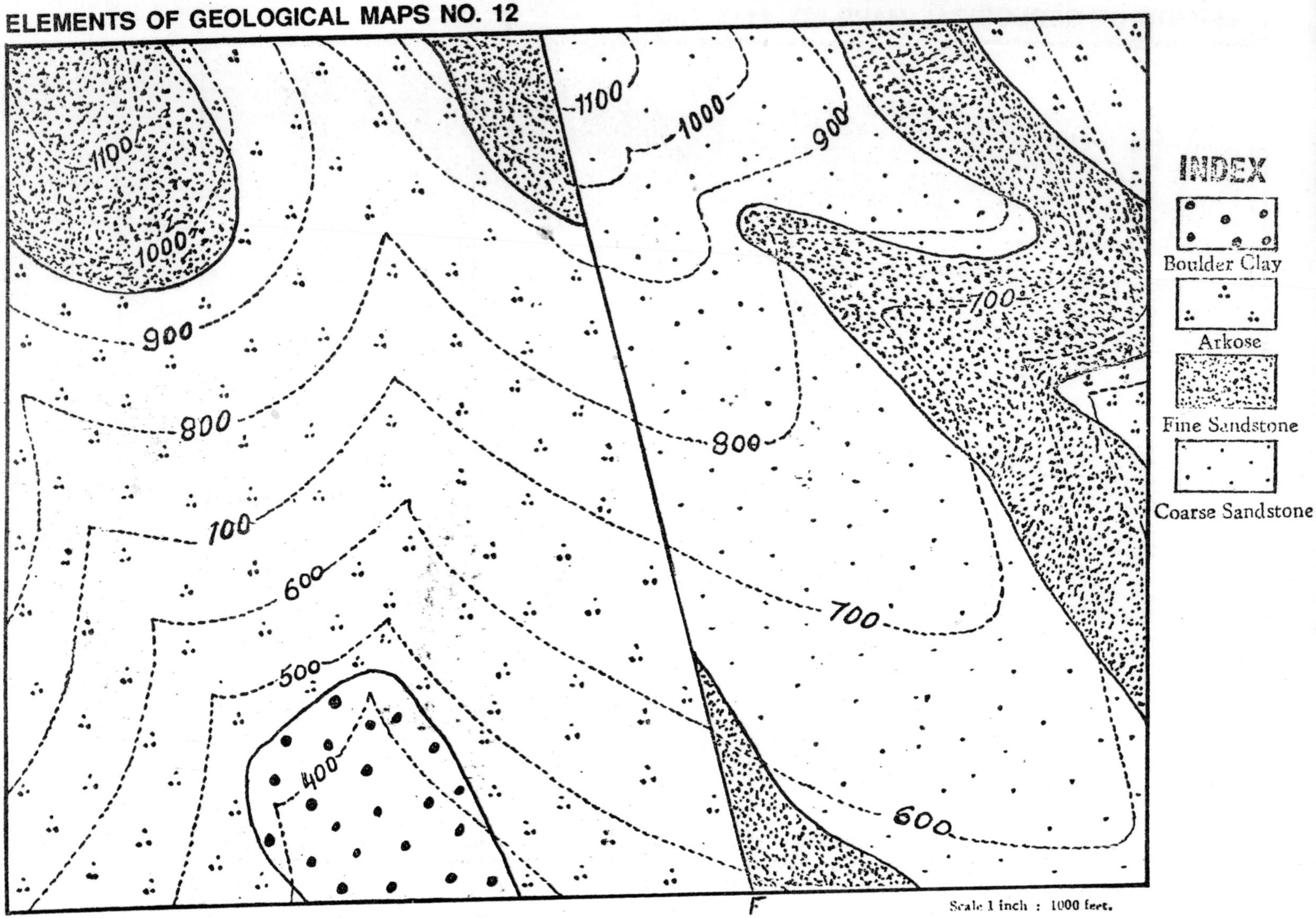

Scale 1 inch : 1000 feet.

Describe the geology of the area. Draw a suitable section to bring out the structure.

ELEMENTS OF GEOLOGICAL MAPS NO. 13

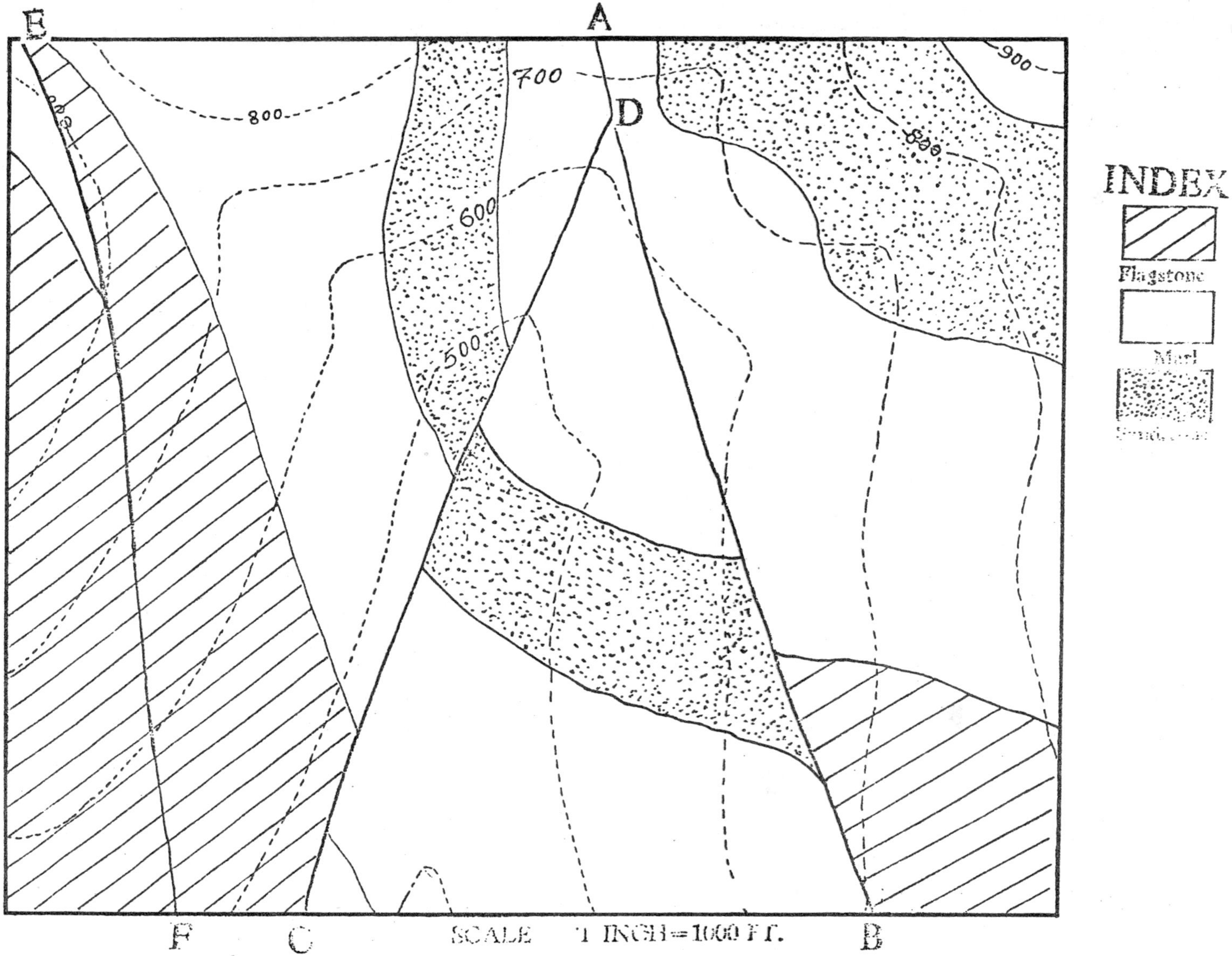

ELEMENTS OF GEOLOGICAL MAPS NO. 14

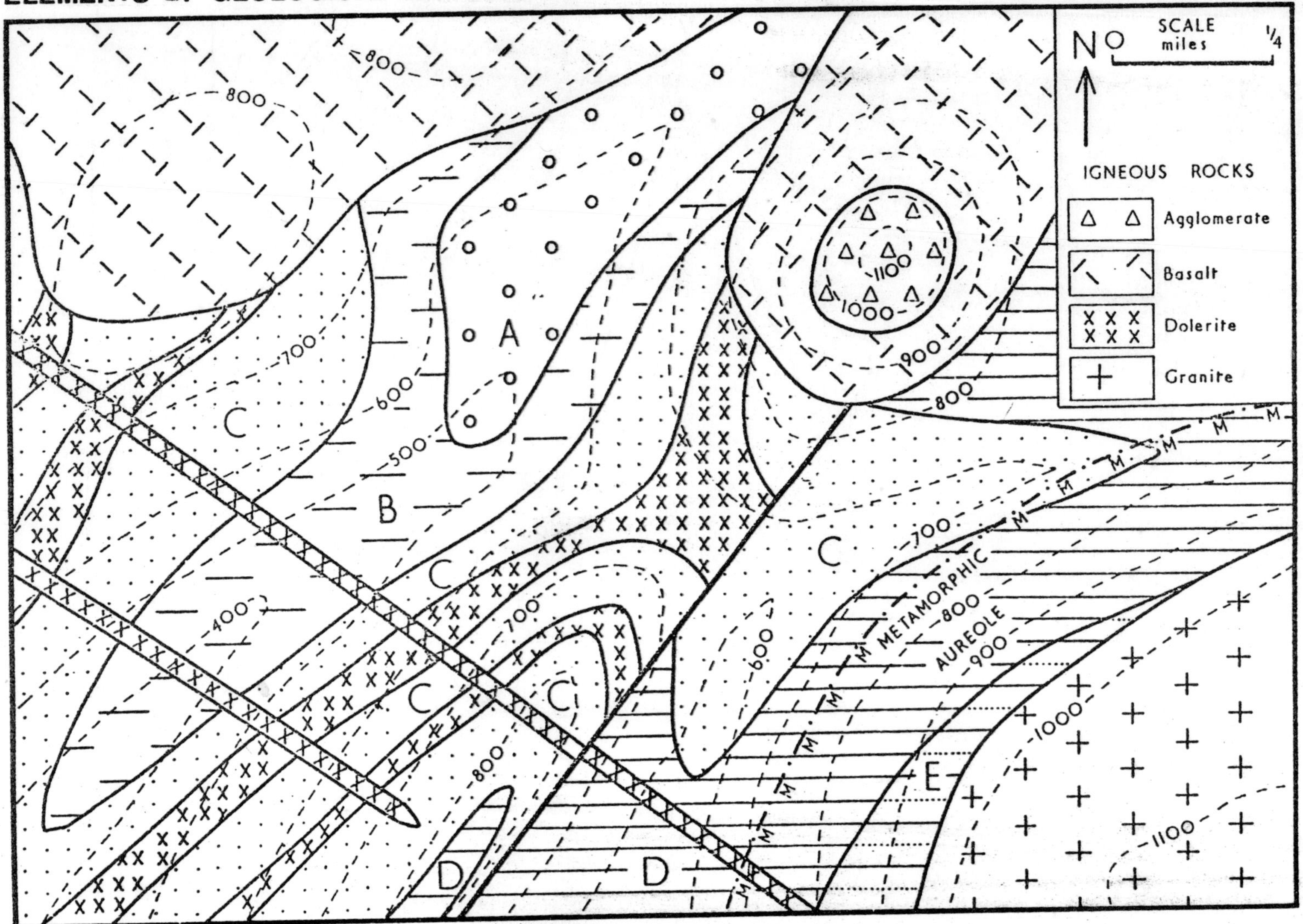

Map [illegible]. Draw sections across the map to show the form of the igneous rocks and the other structural features. Also deduce the relative ages of the igneous rocks as far as this is possible.

ELEMENTS OF GEOLOGICAL MAPS NO. 15

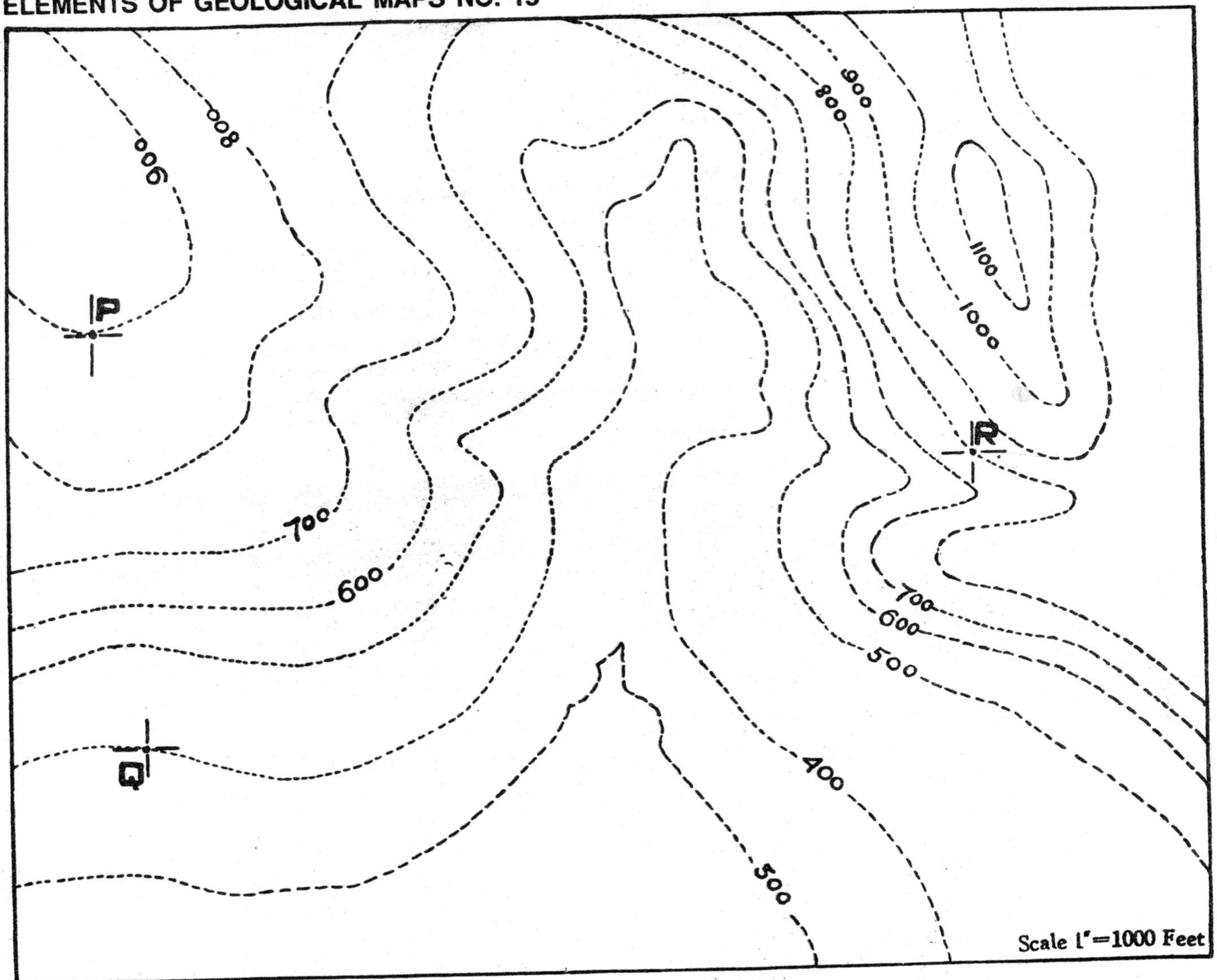

THREE POINTS PROBLEM Three points P, Q and R are given, complete the outcrops assuming the coal-seam to have a constant strike and dip.

ELEMENTS OF GEOLOGICAL MAPS NO. 16

SCALE—1 Inch : 1000 feet

(By Courtesy of M/s. George Allen & Unwin Ltd. London, From the Book "Elementary Exercises upon Geological Maps" by John I-Platt).

ELEMENTS OF GEOLOGICAL MAPS NO. 17

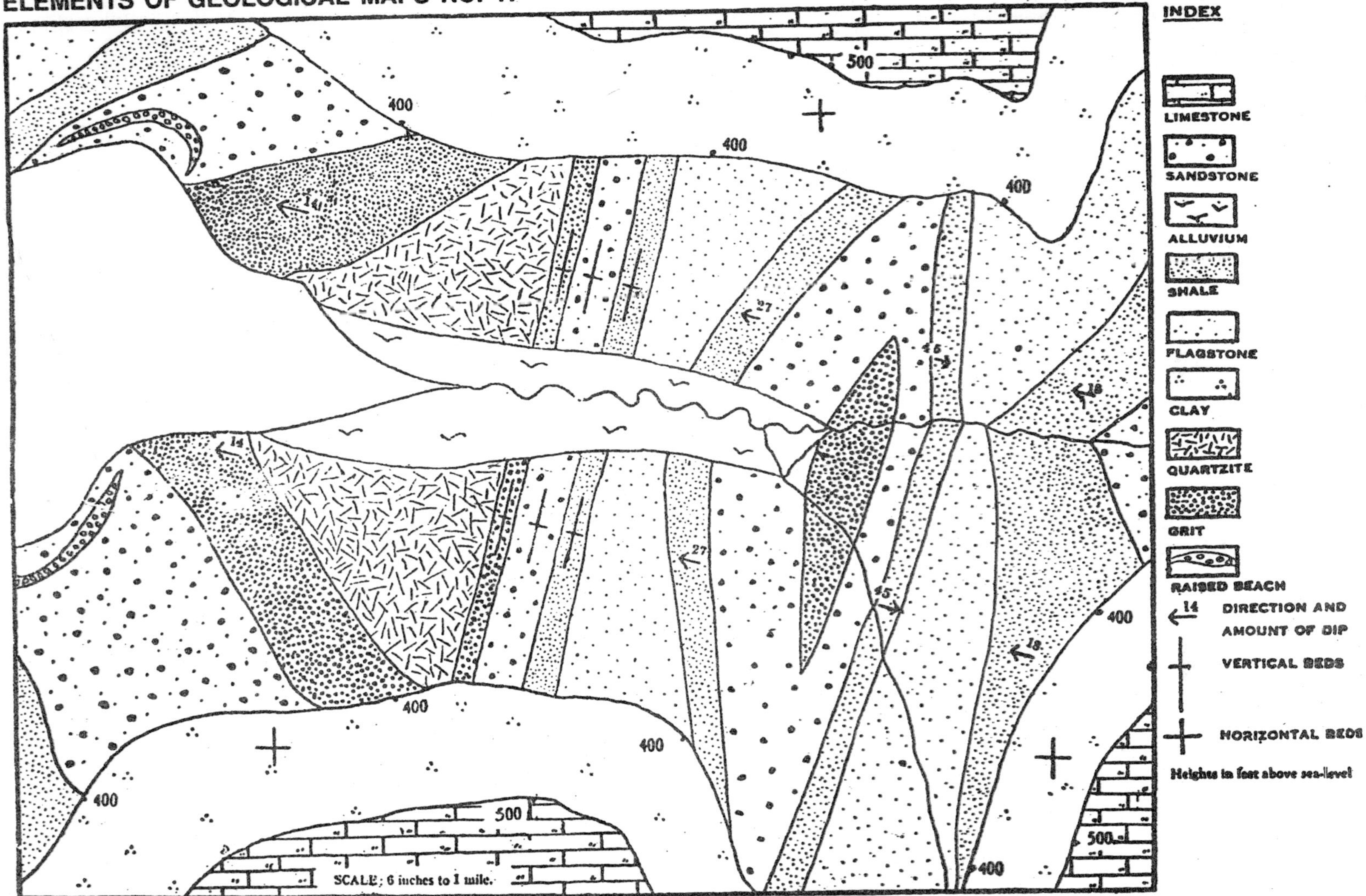

(By Courtesy of M/s. George Allen & Unwin Ltd. London, From the Book "Elementary Exercises upon Geological Maps" by John I-Platt).

NOTE—
In the index to this map, the beds have not been arranged in their proper order.
Part of the exercise is to find out the true sequence.

APPENDICES
(SOME USEFUL TABLES)

APPENDIX 1

Table I: Natural Sines, Cosines, Tangents, and Cotangents

Angle in Degrees	Sine	Tangent	Cosine	Cotangent	Angle in Degrees
1°	·0175	·0175	·9999	57·2900	89°
2°	·0349	·0349	·9994	28·6363	88°
3°	·0523	·0524	·9986	19·0811	87°
4°	·0698	·0699	·9976	14·3007	86°
5°	·0872	·0875	·9962	11·4301	85°
6°	·1045	·1051	·9945	9.5144	84°
7°	·1219	·1228	·9926	8·1444	83°
8°	·1392	·1405	·9903	7·1154	82°
9°	·1564	·1584	·9877	6·3138	81°
10°	·1737	·1763	·9848	5·6713	80°
11°	·1908	·1944	·9816	5·1446	79°
12°	·2079	·2126	·9782	4·7046	78°
13°	·2250	·2309	·9744	4·3315	77°
14°	·2419	·2493	·9703	4·0108	76°
15°	·2588	·2680	·9659	3·7321	75°
16°	·2756	·2868	·9613	3·4874	74°
17°	·2924	·3057	·9563	3·2709	73°
18°	·3090	·3249	·9511	3·0777	72°
19°	·3256	·3443	·9455	2·9042	71°
20°	·3420	·3640	·9397	2·7475	70°
21°	·3584	.3839	·9336	2·6051	69°
22°	·3746	·4040	·9272	2·4751	68°
23°	·3907	·4245	·9205	2·3559	67°
24°	·4067	·4452	·9136	2·2460	66°
25°	·4226	·4663	·9063	2·1445	65°
26°	·4384	·4877	·8988	2·0503	64°
27°	·4540	·5095	·8910	1·9626	63°
28°	·4695	·5317	·8830	1·8807	62°
29°	·4848	·5543	·8746	1·8041	61°
30°	·5000	·5774	·8660	1·7321	60°
31°	·5150	·6009	·8572	1·6643	59°
32°	·5300	·6249	·8480	1·6003	58°
33°	·5446	·6494	·8387	1·5399	57°
34°	·5592	·6745	·8290	1·4826	56°
35°	·5736	·7002	·8192	1·4282	55°
36°	·5878	·7265	·8090	1·3764	54°
37°	·6018	·7536	·7986	1·3270	53°
38°	·6157	·7813	·7880	1·2799	52°
39°	·6293	·8098	·7772	1·2349	51°
40°	·6428	·8391	·7660	1·1918	50°
41°	·6560	·8693	·7547	1·1504	49°
42°	·6691	·9004	·7431	1·1106	48°
43°	·6820	·9325	·7314	1·0724	47°
44°	·6947	·9657	·7193	1·0355	46°
45°	·7071	1·0000	·7071	1·0000	45°
Angle in Degrees	Cosine	Cotangent	Sine	Tangent	Angle in Degrees

APPENDIX 2

Table II: Showing Variation of Dip with Change of Direction (after Jukes)

Angle between directions of true dip and section	Angle of Full Dip																
0°	10°	15°	20°	25°	30°	35°	40°	45°	50°	55°	60°	65°	70°	75°	80°	85°	89°
10°	9° 51′	14° 47′	19° 43′	24° 48′	29° 37′	34° 36′	39° 34′	44° 34′	49° 34′	54° 35′	59° 37′	64° 40′	69° 43′	74° 47′	79° 51′	84° 56′	88° 59′
15°	9° 40′	14° 31′	19° 23′	24° 15′	29° 9′	34° 4′	39° 2′	44° 1′	49° 1′	54° 4′	59° 8′	64° 14′	69° 21′	74° 30′	79° 39′	84° 50′	88° 58′
20°	9° 24′	14° 8′	18° 53′	23° 39′	28° 29′	33° 21′	38° 15′	43° 13′	48° 14′	53° 19′	58° 26′	63° 36′	68° 49′	74° 5′	79° 22′	84° 41′	88° 56′
25°	9° 5′	13° 39′	18° 15′	22° 55′	27° 37′	32° 24′	37° 15′	42° 11′	47° 12′	52° 18′	57° 30′	62° 46′	68° 7′	73° 32′	78° 59′	84° 29′	88° 54′
30°	8° 41′	13° 34′	17° 30′	22° 0′	26° 34′	31° 13′	36° 0′	40° 54′	45° 54′	51° 3′	56° 19′	61° 42′	67° 12′	72° 48′	78° 29′	84° 14′	88° 51′
35°	8° 13′	12° 28′	16° 36′	20° 54′	25° 18′	29° 50′	34° 30′	39° 19′	44° 17′	49° 29′	54° 49′	60° 21′	66° 8′	71° 55′	77° 51′	83° 54′	88° 47′
40°	7° 41′	11° 36′	15° 35′	19° 39′	23° 51′	28° 12′	32° 44′	37° 27′	42° 23′	47° 35′	53° 0′	58° 40′	64° 35′	70° 43′	77° 2′	83° 29′	88° 42′
45°	7° 6′	10° 4′	14° 25′	18° 15′	22° 12′	26° 20′	30° 41′	35° 16′	40° 7′	45° 17′	50° 46′	56° 36′	62° 46′	69° 14′	76° 0′	82° 57′	88° 35′
50°	6° 28′	9° 46′	13° 10′	16° 41′	20° 21′	24° 14′	28° 20′	32° 44′	37° 27′	42° 33′	48° 4′	54° 2′	60° 29′	67° 22′	74° 40′	82° 15′	88° 27′
55°	5° 46′	8° 44′	11° 48′	14° 58′	18° 19′	21° 53′	25° 42′	29° 50′	34° 21′	39° 20′	44° 47′	50° 53′	57° 36′	64° 58′	72° 75′	81° 20′	88° 15′
60°	5° 2′	7° 38′	10° 19′	13° 7′	16° 6′	19° 18′	22° 45′	26° 33′	30° 47′	35° 32′	40° 54′	46° 59′	53° 57′	61° 49′	70° 34′	80° 5′	88° 0′
65°	4° 15′	6° 28′	8° 45′	11° 9′	13° 43′	16° 29′	19° 31′	22° 55′	26° 44′	31° 7′	36° 14′	42° 11′	49° 16′	57° 37′	67° 21′	78° 19′	87° 38′
70°	3° 27′	5° 14′	7° 6′	9° 3′	11° 10′	13° 28′	16° 0′	18° 53′	22° 11′	26° 2′	30° 29′	36° 15′	43° 13′	51° 55′	62° 43′	75° 39′	87° 5′
75°	2° 37′	3° 33′	5° 23′	6° 53′	8° 30′	10° 16′	12° 15′	14° 30′	17° 9′	20° 17′	24° 8′	29° 2′	35° 25′	44° 1′	55° 44′	71° 20′	86° 9′
80°	1° 45′	2° 40′	3° 27′	4° 37′	5° 44′	6° 56′	8° 17′	9° 51′	11° 41′	13° 55′	16° 44′	20° 25′	25° 30′	32° 57′	44° 33′	63° 15′	84° 15′
85°	0° 53′	1° 20′	1° 49′	2° 20′	2° 53′	3° 30′	4° 11′	4° 59′	5° 56′	7° 6′	8° 35′	10° 35′	13° 28′	18° 1′	26° 18′	44° 54′	78° 41′
89°	0° 10′	0° 16′	0° 22′	0° 28′	0° 35′	0° 42′	0° 50′	1° 0′	1° 11′	1° 26′	1° 44′	2° 9′	2° 45′	3° 44′	5° 31′	11° 17′	44° 15′

APPENDIX 3

Table III : Depth to Stratum below Horizontal Surface for Various Distances and Dips (after Hayes)

Dip Angle	100 ft.	200 ft.	300 ft.	400 ft.	500 ft.	1000 ft.	¼ mile or 1320 ft.	½ mile or 2640 ft.	¾ mile or 3960 ft.	1 mile or 5280 ft.
1°	1·75	3·50	5·25	7·00	8·75	17·5	23·4	46·08	69·12	92·16
2°	3·49	6·98	10·47	13·96	17·45	39·4	46·09	92·18	138·3	184·4
3°	5·24	10·48	15·72	20·96	26·20	52·4	69·18	138·4	207·5	276·7
4°	6·99	13·98	20·97	27·96	34·95	69·9	92·30	184·6	276·9	369·2
5°	8·75	17·50	26·25	35·00	43·75	87·5	115·5	230·5	346·8	461·9
6°	10·51	21·02	31·53	42·04	52·55	105·1	138·7	277·4	416·1	555·0
7°	12·28	24·56	36·48	49·12	61·40	122·8	162·1	324·2	486·3	448·3
8°	14·05	28·10	42·15	56·20	70·20	140·5	185·5	371·0	556·5	742·0
9°	15·84	31·68	47·52	63·36	79·20	158·4	209·1	418·2	627·3	836·3
10°	17·63	35·26	52·89	70·52	88·15	176·3	232·8	456·6	698·4	931·0
11°	19·44	38·88	58·32	77·76	97·20	194·4	256·6	513·2	769·8	1023
12°	21·26	42·52	63·78	85·04	106·30	212·6	280·6	561·2	841·8	1126
13°	23·09	46·18	69·27	92·36	115 45	230·9	304·7	609·4	914·1	1219
14°	24·93	49·86	74·79	99·72	124·65	249·3	329·1	658·2	987·3	1316
15°	26·80	53·60	80·40	107·20	134·00	268·0	353·7	707·4	1060	1415
16°	28·68	57·36	86·04	114·72	143·40	286·8	378·5	757·0	1136	1514
17°	30·57	o1·14	91·71	122·28	152·85	305·7	403·6	807·2	1211	1614
18°	32·49	64·98	97·47	129 96	164·45	324·9	428·9	857·8	1287	1716
19°	34·43	68·86	103·29	137·72	172·15	344·3	454·3	908·6	1363	1817
20°	36·40	72·80	109·20	145·60	182·00	364·0	480·4	960·8	1411	1923
21°	38·39	76·78	115·17	153·56	191·95	383·9	506·7	1012	1520	2027
22°	40·40	80·80	121·20	161·60	202·00	404·0	533·3	1067	1600	2133
23°	42·45	84·90	127·35	169·80	212·25	424·8	560·3	1121	1681	2241
24°	44·52	89 04	133·56	178·08	222·60	455·2	587·7	1175	1763	2351
25°	46·63	93·26	139·89	186·52	233·15	466·3	615·5	1231	1847	2462
26°	48·77	97·54	146·31	195·08	243·85	487·7	643·7	1287	1931	2575
27°	50·95	101·90	152·85	203·80	254·75	509·5	672·6	1345	2018	2690
28°	53·17	106·34	159·51	212·68	265·85	531·7	701·8	1404	2105	2807
29°	55·43	110·86	166·29	221·72	277·15	554·3	731·7	1463	2195	2927
30°	57·74	115·48	173·22	230·96	288·70	577·4	762·1	1524	2286	3048